Economic Analysis of Investment Projects

ECONOMIC ANALYSIS OF INVESTMENT PROJECTS

A Practical Approach

Kedar N. Kohli

Published for the Asian Development Bank
Oxford University Press

Oxford University Press, Hong Kong

Oxford New York Toronto
Delhi Bombay Calcutta Madras Karachi
Kuala Lumpur Singapore Hong Kong Tokyo
Nairobi Dar es Salaam Cape Town
Melbourne Auckland Madrid

and associated companies in
Berlin Ibadan

Oxford is a trade mark of Oxford University Press

First published 1993

Published in the United States
by Oxford University Press, New York

British Library Cataloguing in Publication Data
available

Library of Congress Cataloging in Publication Data
available

ISBN 0 19 585936 7 (HB)
ISBN 0 19 585937 5 (PB)

Printed in Hong Kong
Published by Oxford University Press
18/F Warwick House, Tong Chong Street, Quarry Bay, Hong Kong

Contents

List of Tables

Figure

Preface

The economic analysis of projects is considered by many as an esoteric subject, understood only by a select few, mostly economists who have a good knowledge of mathematics. This observation is not without basis because while several books have been written on the subject since the 1960s, there are very few books which can be easily understood by noneconomists, and even by a majority of economists. In writing this book, I, therefore, kept two objectives in mind: The first objective is to explain in nontechnical terms the meaning, purpose, and methodology of economic analysis, and how such an analysis differs from the financial analysis of projects. The economic analysis is supported by simple examples which can be easily understood by economists and noneconomists alike. The second objective is to focus attention on areas which are essential in making a sound assessment of the economic—as distinct from the financial—viability of projects. Both objectives take into consideration the data limitations which exist in most developing countries.

In preparing this book, I have, with the approval of the Management of the Asian Development Bank (ADB), made extensive use of ADB's *Guidelines for the Economic Analysis of Projects*. These *Guidelines* were finalized by me in 1983, with the help of several other economists in ADB. However, I have extensively rewritten the text and added several new chapters. Consequently, this book differs greatly from the 1983 publication. Furthermore, the focus of the two documents is very different. While the *Guidelines* constitute staff instructions, this book is intended for wider use by students of economics and by professionals in developing countries. It can also be used by bilateral and multilateral aid institutions that deal with project formulation, appraisal, and evaluation.

I am grateful to the management of ADB for allowing the Bank to publish this book. My special thanks to Mr. George Liu, Chief Information Officer, and Dr. A.I. Aminul Islam, Chief of ADB's Development Policy Office, for their continued encouragement and support in preparing this book for publication. In particular, I would like to express my deep gratitude to Dr. Ifzal Ali, ADB Senior Economist, who extensively reviewed the entire draft more than once and has made very valuable comments which have been duly incorporated into the book. Finally, I am greatly indebted to Ms. Cora O. Faustino, Ms. Marinie I. Baguisa, and my wife Mrs. Sudesh Kohli for their unstinting assistance in typing the manuscript. I, however, remain fully responsible for any mistakes or errors that may still be found in the book.

KEDAR N. KOHLI

Economic Analysis of Investment Projects
A Practical Approach

CHAPTER 1

Objectives and Scope

Introduction

This book attempts to provide a simple, step-by-step approach for the economic analysis of projects. This subject has been an active field of research and study since the early 1960s and many articles, textbooks, and guidelines have been written on it. However, many of the textbooks and articles are written in technical language which can be understood only by economists who have a good knowledge of mathematics. The aim of this book is to explain the methodology of the economic analysis of projects in a language which can be understood not only by economists but also by engineers, financial analysts, bureaucrats, and other persons involved in project appraisal.

The theoretical foundations of the economic or social analysis of projects—on which a great deal of work has already been done—will not be discussed in detail. Instead, it will be demonstrated through simple illustrations how economic analysis can be applied in realistic situations prevailing in most developing countries. It should be recognized that these countries, in general, face data limitations and manpower constraints. Therefore, any methodology which demands a great deal of data or skilled manpower is not likely to receive wide acceptance.

The method of analysis greatly depends on the national objectives sought to be achieved through the implementation of projects. This issue has been a subject of considerable discussion and debate since the 1960s. Initially, economists had considered the maximization of national consumption or output as the principal development objective. However, with the growing concern about the inadequacy of public savings and problems of poverty during the 1970s, the emphasis shifted to social cost-benefit analysis in which weights were assigned to additional consumption and savings resulting from the use of project inputs and outputs.

Although a great deal of resources were spent by international institutions on the development of social cost-benefit analysis, its practical use has remained extremely limited even among those institutions. The reasons for this lie primarily in the complexity of analysis and the difficult value judgment required on the weights to be attached to the income accruing to the beneficiaries.

In this book, the maximization of national output is considered as the primary objective of economic development. The test of a project's viability, therefore, depends upon its net contribution to national output, often referred to as the "economic" or "efficiency" objective. In this analysis, no distinction is made as to whom the income accrues and the purpose for which it is used. The economic prices are calculated on the assumption that an extra unit of consumption is as valuable as an extra unit of investment. It is assumed that if a certain government wishes to raise investment resources, it is best done through fiscal means, while income distribution and employment objectives can best be realized through the appropriate design and location of projects.

Objectives of Development

One of the most notable events of postwar economic history is the political independence of most developing countries and their deliberate effort to promote economic development through the formulation and implementation of development plans and programs which extend over several years. While each country has its own socioeconomic priorities, all hold certain objectives in common. The primary objective of virtually all development programs is to achieve rapid and sustained economic growth—consistent with the aim of providing gainful employment to a growing labor force and fulfilling the basic needs of all citizens.

In formulating their development plans, most developing countries are faced with the basic economic problem of allocating limited national resources of capital, skilled manpower, land, and foreign exchange to meet the diverse needs of the economy. The ultimate objective of this allocative process is to maximize economic growth and thereby improve the well-being of all members of a nation's society. Therefore, a choice has to be made among competing uses of resources in a manner that maximizes the national objectives. This requires the determination of a suitable set of criteria which will help in making a rational and optimum allocation of resources.

The formulation of a national development plan involves three distinct but interrelated stages. The first stage involves the preparation of a macroeconomic development plan which provides a consistent framework on the relationship between the availability of investment resources—including those financed through external assistance—and the physical program of various sectors on which the overall growth targets are based. In formulating the macroeconomic plan covering several years, an appropriate strategy is designed to make full use of all available resources and to remove any bottlenecks that might impede growth in the country. For instance, if foreign exchange resources are extremely scarce and unskilled labor is abundant, a greater stress may be placed on those sectors which make limited claim on external resources and employ labor-intensive technology. Similarly, if shortage of skilled labor is a major constraint in implementing and operating projects, special stress may be put on expanding and improving technical education.

The macroeconomic development plan is normally prepared for the entire economy. However, for purposes of implementation, a distinction is made between the public and the private sectors. While the private sector plan is generally indicative in nature, the public sector plan is spelled out in detail. It is through the sectoral plans that the overall development plan of the public sector is implemented. In preparing the sectoral plans, various issues such as physical targets, their internal consistency, choice of techniques, institutional arrangements, time phasing, and sources of financing for implementing these plans need to be indicated. In determining the sources of financing, the pricing policies of goods and services provided by each sector and the basis of determining them assume crucial importance.

The third stage in the formulation of the public sector development plan is to identify, prepare, and implement specific investment projects to achieve sectoral objectives. Projects are thus the building blocks of the public sector development plans, since all such plans are implemented through specific projects. The availability of well-prepared projects to match the investment funds provided for each sector is important to ensure consistency between investment outlays and physical programs and for the timely realization of sectoral objectives.

Relevance of Project Analysis

In many developing countries, direct and indirect government investments constitute an important part of the total national development program. For instance, in practically all developing countries, governments are responsible for investment in social sectors like education, health, and water supply, and in economic infrastructure like transport and electricity. In addition, in many countries the public sector continues to play a major role in agriculture and industry. Therefore, the performance of public sector projects has a crucial bearing on the overall performance of the national economy.

Given the importance of the public sector, it is essential to formulate a suitable set of criteria before public funds are invested so that their contribution to the national economy can be maximized. The private or financial criterion, in which the return on an investor's equity provides the primary basis for the choice of investment, is not particularly relevant since all funds to public enterprises are provided by the government and loan funds generally carry subsidized rates. Also, as will be explained later, some of the items included in the cost and benefit streams are not relevant in determining the profitability of public funds. More importantly, private tests of profitability cannot be directly applied, either because market prices for some of the public goods do not exist or because these prices are not sound indicators of the value of goods and services provided by the government. For instance, roads do not carry tolls, irrigation projects do not attract water charges, and drinking water is supplied at highly subsidized rates. Furthermore, because of various distortions, market prices do not often provide a correct basis for determining the real cost of inputs used and output produced by the project.

Cost-benefit analysis was developed as a method for determining the economic soundness and viability of projects. This analysis was used in a few instances in developed countries before the Second World War, but its systematic use for project analysis began only in the 1960s. The concern for sound investment criteria for investment projects initially arose among international lending institutions whose assistance was sought by developing countries for financing investment projects in various sectors.

In economic analysis, benefits are defined in terms of their effect on national output, while costs are defined in terms of their opportunity cost, i.e., the benefits foregone by not using the resources in the next best available alternative. Defining the benefits and costs in this fashion implies that there is no alternative use of resources consumed by the project which would have yielded a higher contribution to national output.

The basic thrust of economic analysis is thus directed on the consideration that, when projects are taken up for implementation, they do not only confer benefits but also involve real costs—since resources withdrawn for use in a project will no longer be available to the rest of the economy. These benefits and costs must be properly compared and only those projects which maximize the net real gains—benefits minus costs—should be taken up for implementation.

It should be noted that the excess of benefits over costs is by itself not an adequate basis for implementing a project. In practice, there may be several alternatives in terms of technology, design, location, or time phasing for achieving the objectives of the project. For instance, coal for power plants may be moved by railways, road, or conveyor system. Power itself may be generated from various sources such as hydro, coal, gas, etc. Various available options must be compared and the one which involves the least cost and at the same time generates the highest surplus should be selected. This consideration of alternatives is the single most important feature of project analysis. To ensure that the economic analysis makes maximum contribution in the efficient use of resources, such consideration must be taken into account at an early stage of the project cycle.

In determining the economic costs and benefits of a project, it is essential to state the objectives clearly. This will help define the project boundary, and all costs required to achieve the project's objectives must be included in the project's economic cost. For instance, if the objective of a power project is to increase the supply of electricity by a given amount in an area, then all costs of generation, transmission, and distribution necessary to achieve that objective must be included to determine the viability of the project. This may not be done in financial analysis if, for instance, the power supply company is only concerned with the generation of power and the responsibility for transmission and distribution rests with another company.

Cost-benefit analysis is thus a method of making the choice between the competing use of resources in a consistent and comprehensive fashion. In essence, it involves estimation of the real—as distinct from financial—benefits and costs of a project to the economy, reducing them to a common denominator, and comparing the flows of benefits and costs which accrue at different time periods. Only when benefits exceed costs may the project be accepted. In certain situations, however, the policy makers may decide to implement a project whose return is below an agreed norm of project viability. This could arise out of social considerations such as regional development or development of backward areas. In such cases, economic analysis helps policy makers know the real economic cost involved in implementing such decisions.

The main task of an economic project analyst is thus twofold. The first is to select the least-cost option among the available, feasible alternatives. The second is to identify real economic costs and benefits of the selected project, express them in monetary value, measure them with a common yardstick, and compare them to determine the net impact of the project. These issues are dealt with in subsequent chapters.

Economic analysis does not end with the appraisal stage of the project. It is possible that several underlying assumptions may change during the implementation phase which may necessitate changes in the scope of the project. Therefore, economic analysis may be required during project implementation to ensure continuing viability. After project completion, *ex-post* evaluation is essential so that lessons learned can be incorporated in the design and implementation of future projects. Thus, economic analysis is important at all stages of the project cycle.

It may be mentioned that public sector projects, for which cost-benefit analysis has been specifically developed, generally have long lives, ranging from 15 to 50 years. During such extended periods, many changes could occur that may alter the value or relevance of project benefits. These changes may greatly affect the relationship between costs and benefits assumed during appraisal and thereby alter the viability of the project. This is, however, a risk inherent in the nature of public sector projects and does not constitute any adverse reflection on the cost-benefit methodology itself.

Financial versus Economic Analysis

Economic analysis of projects is somewhat similar to financial analysis because both assess the profitability of investment. But the concept of financial profit is not the same as economic profit. While economic profitability indicates the real worth of a project to the country, financial profitability provides a measure of the commercial viability of the project. In making public investment decisions, both financial and economic profitability should be considered. It has sometimes been suggested that financial profitability not be made a concern because as long as a project is economically sound, it can be supported through government subsidies. In practice, this rarely happens as most governments face severe budgetary constraints. Consequently, the affected project entity may run into severe liquidity problems, thereby jeopardizing even its economic viability.

Financial analysis is concerned with estimating the rate of return of a project to the project entity based on the prevailing market prices of inputs used and outputs produced by the project. The project authority has to determine the quantity and mix of products which will maximize the return on total investment. As long as the return on total investment exceeds the market interest rate, the project entity will attempt to borrow as much funds as possible to increase the return on its own invested funds.

Economic analysis, on the other hand, is concerned with estimating the return on investment to the national economy, as distinct from the project entity. The objective of this analysis is to suggest a methodology which will help a project maximize the return on total investment to the national economy. Since all funds for the project are provided by the government, a distinction between the sources of funds is not relevant. Very often a comparison is made between the financial and economic rates of return on the total capital employed in a project. The two returns are not strictly comparable. An examination of the reasons for this difference provides the justification of the economic—as distinct from financial—analysis of projects.

In financial analysis all costs incurred by the project entity at prevailing market prices are included in estimating total cost, while the benefits are equal to money income received by the project entity. In economic analysis, on the other hand, costs include only those items in the project which impose a real burden on the economy, while benefits are measured by the extent to which the project adds to the national output. In economic analysis, all financial costs and benefits, which merely represent a transfer from one sector to another, are excluded. On the other hand, if some additional investment on and/or expenditure of real resources elsewhere in the economy are necessary to realize the full benefits of the project, they must be included in the economic cost of the project.

In most developing countries, market prices of goods, services, capital, and foreign exchange are distorted for a variety of reasons. They do not serve as a meaningful basis for determining the economic cost of inputs used and value of output produced by the project. For instance, many developing countries suffer from a large degree of unemployment. In this situation, the real cost to society of withdrawing labor from other sectors for use in the project will be far below the financial cost. Similarly, market prices do not provide a good measure of the benefits if there is no freely operating market—as in water supply or electricity—for its output. In this situation, the willingness to pay on the part of the consumers provides the economic measure of the benefits of the project. Valuation of the economic benefits and costs is, therefore, the single most important part of the economic analysis of projects.

In some instances, the existence or operation of a project may confer some benefits or impose some costs on the economy. It could be that such costs or benefits are not included in the financial accounts of the project. They are known as external economies or diseconomies of a project. For instance, untreated discharge from a fertilizer plant may increase its financial profit but would add to the production cost of downstream industries. The cost of the pollution treatment plant must be included in determining the economic viability of the fertilizer project.

A project may use inputs or produce outputs which may either be exported or imported (traded goods) or which are meant entirely for domestic use. In financial analysis, the use of market prices makes this valuation simple. In economic analysis, the traded and nontraded goods, for reasons that will be explained later, are valued separately and a common denominator (called numeraire) has to be found for combining the two components.

Two kinds of numeraire are commonly used for this purpose: the aggregate consumption or willingness to pay (WTP) numeraire and the foreign exchange numeraire. The WTP numeraire values nontraded goods and services based on what society is willing to pay. It first values traded goods and services in foreign exchange at border prices and then converts them into local currency through the use of a shadow exchange rate to make them additive with nontraded values. The foreign exchange numeraire reverses this process. Traded goods are valued directly in terms of their direct effect on foreign exchange at border prices, while nontraded goods are valued in terms of their indirect effect on foreign exchange. Commodity, group, and/or standard conversion factors are used to trace the foreign exchange effect of nontraded goods. In this book, the foreign exchange (or convertible currency) numeraire is used, and all costs and benefits are expressed in border prices but in units of local currency at the official exchange rate.

In short, the essential difference between financial and economic analysis is that while the former is concerned with financial profitability based on market prices, the latter is aimed at maximizing the net impact of investment on the national economy. Economic analysis is thus concerned with the overall profitability of a project based on its real economic costs and benefits. Hence, such a project's benefits and costs may differ from market prices depending on the degree of distortion prevailing in the market. Other differences may also arise from some of the economic costs and benefits which may not have been taken into account in the determination of the financial profitability of the project. The purpose of economic analysis is to suggest a sound and practical methodology for determining the real impact of a project on national output so that an informed decision can be taken about its implementation.

Determination of Project Viability

In both the financial and economic analysis of a project, a comparison has to be made between the flows of costs and benefits to determine its profitability. Since these flows take place at different time periods, they cannot be directly compared or added up. This is because waiting involves a cost; a dollar earned today is not equal to a dollar earned two years later. The streams of costs and benefits must be discounted to assess the financial or economic viability of a project.

The discount rate that makes the flows of costs and benefits equal provides a measure of project profitability. In financial analysis, the discount rate is called the financial internal rate of return (FIRR), while in economic analysis, it is known as the economic internal rate of return (EIRR).

Two points may be noted in this context. First, both FIRR and EIRR represent the return on total resources used for investment and production. In financial analysis, if interest on borrowed funds is below the FIRR, the return on equity (before taxes) will exceed the interest on borrowed funds, and the higher the difference, the better the project from the owner's viewpoint. In economic analysis, the EIRR is compared with the opportunity cost of capital (OCC), or what the economy can earn in alternative projects. If the EIRR exceeds the OCC, the project is considered viable. Second, since both financial and economic costs and benefits are expressed in constant prices, interest on borrowed funds and OCC—the benchmark for profitability—should also be measured in constant prices.

In addition to the EIRR, other tools can be used in determining the economic viability of projects. While these tools will be discussed in detail later on, the most important one, the net present value (NPV) will be briefly referred to in this chapter. Under the NPV, the flows of costs and benefits are discounted at a rate which is equal to the OCC. If benefits exceed costs, the NPV becomes positive and the project should be implemented. If, however, NPV is negative the project should be rejected. A negative PNV means that the project would use more real resources than it would produce over time. Since the present value of the flow of costs and benefits greatly varies with the discount rate used, the choice of that rate (OCC) is crucial in the NPV.

Limitations of Economic Analysis

Cost-benefit analysis was introduced in the 1960s by donor countries and institutions essentially for two reasons. First, the markets prevailing in developing countries during that time were highly distorted and did not provide the correct signals for making economically sound investment choices. It was, therefore, felt that the use of market prices in making investment decisions will perpetuate those distortions and even aggravate them further. Second, since market prices did not reflect the relative scarcity of resources in the country, their use would lead to a misallocation of resources.

In cost-benefit analysis it is assumed that market distortions will continue. Shadow prices, representing the opportunity cost of resources used and outputs produced, are used to determine the economic viability of individual projects. An implicit assumption is made that the existing relationship between market and economic prices will continue throughout the life of the project. Such an analysis reflected both the prevailing strong interventionist role of governments in developing countries during the 1960s and 1970s and the pessimism among economists about governments' ability to implement major policy changes to remove the distortions.

One of the major criticisms of cost-benefit analysis has been the implicit assumption that as long as a project is economically sound, financial considerations need not come in the way of its implementation. For instance, if a fertilizer project was found viable based on the international prices of inputs and outputs, the low domestic price of fertilizer fixed by the government could be ignored. After all, the government, in such a scenario, could always subsidize the project. But here, sufficient consideration was not given to the fact that the government may not have the capacity to subsidize such diverse activities like agriculture, power generation, roads, ports, education, and health.

In recent years, it has been recognized that even economically sound projects cannot succeed in unfavorable policy environments. The emphasis has, therefore, shifted from project appraisal *per se* to policy reforms, both at the macro and sectoral level. At the macro level, high import duties and overvalued exchange rates reduce industries' incentive to improve efficiency and productivity and be competitive with imported goods. These policies have also discouraged export promotion. As a consequence, many countries have faced a growing scarcity of foreign exchange which has led to imported raw material shortages. Such shortages affect the operations and profitability of public sector enterprises. Low or below-cost sale prices fixed by governments for various products have added further to the problems of public enterprises because they leave inadequate resources for the rehabilitation and the improvement of existing facilities.

Most developing countries are now more aware that public sector enterprises have not performed well. Instead of becoming a source of funds for future investment, public firms have burdened the shrinking development resources of governments. In fact, it is now well accepted that unless the performance of public firms is improved through institutional and policy reforms, the financial resource problems of developing countries will worsen further. Special emphasis is, therefore, being placed on macroeconomic and sectoral policy reforms. The major areas of reform include: the establishment of realistic exchange rates, greater fiscal discipline, decentralization of decision making, removal of price controls, and improving the efficiency and productivity of enterprises.

Macroeconomic and sectoral policy reforms will not only help reduce the distortions in the prices of goods and services and foreign exchange prevailing in developing countries but will also improve the financial and economic performance of public sector enterprises and reduce their dependence on government funds as a source of financing for future investments. At the same time, economic analysis should help identify the distortions that may exist in specific sectors and the nature of reforms that should be undertaken to improve sectoral performance.

Ideally, if all the macroeconomic and sectoral distortions are removed, the economic and financial prices should become equal, and it may not be necessary to separately carry out the economic analysis of projects. However, in actual practice, many of the distortions would continue to exist in developing countries, and economic analysis will remain an important tool for determining the economic viability of public sector projects. However, greater attention will have to be given than hitherto to the pricing of goods and services and to other institutional aspects so that projects are run efficiently and their dependence on government funds is minimized, if not totally eliminated.

Outline of the Study

This chapter provides an overview of the economic analysis of projects, its relevance and purpose, and how it differs from financial analysis. In subsequent chapters, the various steps involved in carrying out the cost-benefit analysis are explained. Two specific examples are provided to show how this analysis should be undertaken. Chapter 2 gives an explanation of the project cycle, or the stages through which a project passes. This helps in understanding the relevance of economic analysis at each stage and provides a backdrop against which project analysis may be viewed. Chapter 3 briefly describes the method of undertaking demand and least-cost analysis. This analysis is essential to ensure that limited public sector resources are put to the most efficient and effective use. Chapter 4 deals with the method of identifying and quantifying the economic costs and benefits of a project. The chapter starts with the presentation of financial accounting and, thereafter, explains various adjustments that are required to arrive at the economic cost and benefits of a project.

Chapters 5, 6, 7, and 8 deal with the different aspects of economic costs and benefits valuation. Chapter 5 shows the method of determining the value of traded and nontraded goods. Chapter 6 deals with the valuation of labor cost in domestic prices, after taking into account distortions in the labor market. Chapter 7 discusses price changes—particularly relative price changes—an issue that has not been adequately dealt with in most books. Chapter 8 examines the technique of converting the domestic economic values of nontraded goods, services, and labor into border prices so that all costs and benefits can be expressed in border price equivalents.

The final step involves a comparison of economic costs and benefits so that the economic soundness of a project can be determined. Four tests are commonly employed, all involving the determination of the net present value of costs and benefits. This requires the selection of an appropriate national discount rate for each country. Chapter 9 explains the various issues involved in determining the economic viability of a project.

The viability of a project is ascertained by using the most probable values of the parameters in its cost and benefit streams. These values can undergo changes for reasons that are difficult to anticipate. The problem is particularly crucial on the side of benefits that could be spread over several decades. Chapter 10 provides the methodology of analyzing such risks and offers steps that may minimize them.

Chapter 11 deals with benefit monitoring and *ex-post* evaluation. At the appraisal stage, cost-benefit analysis constitutes an ex-ante assessment of the viability of a project. Once it has been decided to proceed with a project, it is essential to ensure that it is implemented without delay and that benefits are actually realized. This chapter examines the various issues that need to be addressed in monitoring the benefits and in making an *ex-post* evaluation.

Chapter 12 describes briefly the social cost-benefit analysis and how it differs from economic analysis. This is to introduce the reader to the nature and complexity of the analysis involved and to explain why its use has not gained wide acceptance in project appraisal. Those who are interested in studying the SCBA further should read the pioneering studies by Little and Mirrlees (1974) and Squire and van der Tak (1975).

Chapters 13 and 14 provide two practical examples of how to use cost-benefit analysis. Chapter 13 illustrates, through the use of actual data, the steps involved in determining the economic viability of an irrigation project where most of the inputs used and outputs produced are internationally traded. Chapter 14 gives an example of a power project whose output is nontraded. These two examples cover most of the issues that are commonly encountered in economic appraisal and evaluation of projects.

The concluding chapter brings together the results of the discussion of the previous chapters. It highlights the distinction between financial and economic measures of profitability and focuses on the future role of cost-benefit analysis in the context of the declining role of the state in investment in developing countries.

CHAPTER 2

Stages in a Project Cycle

A country's development plan must be broken down into specific projects for implementation. The success in achieving development objectives greatly depends upon the thoroughness with which projects are identified, appraised, and assessed for their financial and economic soundness and the efficiency with which they are implemented. Very often there is pressure from government departments, politicians, or local organizations to include projects in the plan whose total cost far exceeds the available resources. Some projects are merely vague ideas which have not been tested for their soundness and viability. As a result, projects get seriously delayed or are poorly implemented, thereby adversely affecting the overall performance of the economy. To maximize their development impact, only those projects which (1) have been thoroughly appraised, (2) make a maximum contribution to national output, and (3) can be fitted into the available development budget should be taken up for implementation.

The work on a project comprises several distinct stages, commonly referred to as the project cycle. The stages of a cycle are closely interlinked and follow a logical sequence. There are four principal stages in the project cycle: identification, appraisal, implementation, and evaluation. The success of a project greatly depends upon how well the work at each stage is carried out. The four stages of the project cycle are briefly discussed below.

Project Identification

The first stage in the project cycle is the identification of a project, which *prima facie* represents a high priority use of a country's resources to achieve the development objectives. At this early stage, it is important to consider a range of alternative approaches to achieve the objectives of the project. This identification and preliminary screening of ideas is an important part of project processing. Decisions made at this stage have a critical and far-reaching impact on the outcome of the project.

For the purpose of such screening, it is essential to ensure that the means chosen to achieve the objectives of the project are cost-effective. At the same time the objectives themselves must be clearly defined for the purpose of determining the size and scope of the project. This requires that all parties concerned with project implementation—including the concerned ministry, ministries of finance and planning, and external aid agency, if any—must agree with the project objectives and strategies. The intended beneficiaries of the project must also be consulted, especially when their participation is essential in ensuring its success, as is the case with projects in the agricultural sector. Failure to reach an agreement about the objectives and to secure firm commitment from all concerned could lead to delays in project implementation, cost over-runs, and nonrealization of the full benefits of the project.

For determining the appropriate size of the project, it is essential to make a realistic assessment of the future demand for the goods and services produced by the project. Without a sound demand analysis, the size of the project may turn out to be either too large or too small, causing either excesses or shortages. A small investment made in making a realistic demand forecast can save a good deal of potential waste of public investment funds. There are several approaches for determining the demand relationship but the most commonly used is the time-series correlation analysis which uses historical data to show how demand varies with price, national output, population, and other relevant factors.

Estimating market demand is particularly important in the case of public utilities like power, roads, and water supply. The growth in their demand can be estimated by studying the relationship over time of the product demand with respect to variables such as national income, industrial production, growth in population, and relative prices. However, it is important to note that in the case of public utilities, government development strategy can itself greatly influence the demand for project output. For instance, growth in demand for electricity can decrease sharply if the emphasis on industrial production is shifted from capital and energy-intensive activities to light consumer goods industries or if prices are increased to reflect the marginal cost of producing electricity. These policies must be clearly articulated and reflected in the feasibility report of the project.

A project may be considered to have passed the identification stage when:

- Major options and alternatives have been identified and some initial choices have been made.
- The principal policy issues affecting the project outcome have been identified and appear amenable to solution.
- The objectives of the project are clearly defined and have the necessary support both from political authorities and from the intended beneficiaries.
- The size of the project is determined based on a realistic assessment of future demand for its output.

Feasibility Study

Having assessed the need for the project, the next step is to determine its feasibility. As the term implies, the purpose is to establish the feasibility or the justification of the project both in its totality as well as its principal dimensions (technical, financial, and economic). A good feasibility report can, therefore, be prepared only through a close interaction among engineers, technical specialists, financial analysts, and economists. At the completion of the feasibility stage, a decision is generally taken on whether to proceed with the project or to drop it. Therefore, it is essential that all aspects of the project are examined objectively and in adequate detail so that an informed decision can be taken by the government.

It is, however, not enough to know that the project yields an adequate rate of return but that the project, as proposed, represents the best available alternative that will maximize its contribution to national welfare. This is done through a series of approximations to test, for instance, different technical approaches for their economic benefits and financial viability. The purpose of this exercise is not to determine that a particular project is good enough to proceed with but to arrive at the best possible one. In arriving at this decision, economic prices should be used for all cost and benefit components of the project. The economist must work closely with the engineer and technical experts to examine the mutually exclusive options and to determine the optimum choice for which a detailed feasibility study should be undertaken.

A well-prepared project also helps minimize the difficulties that might arise in its implementation by anticipating the problems and suggesting measures on how to deal with them. A good feasibility report should include a detailed engineering design, firm cost estimates, institutional arrangements and staffing, and a detailed financial plan for project implementation. Although these details may somewhat delay the preparation of the feasibility report, they would greatly save the time and resources required for project implementation.

The experience of various international agencies in implementing projects shows that there are two major shortcomings in the project preparation stage. In general, the investment cost and the time required to implement the project are underestimated. In some ways these problems are interrelated. The reasons and the extent of underestimation, however, greatly vary with countries and projects. A good feasibility study should anticipate the problems that are likely to arise in implementing a project.

Perhaps the most difficult issue at this stage of project design is to determine the number of components that should be included in a project. There is a golden rule that can be prescribed for this purpose. Much depends upon the institutional and coordination capability of authorities involved in project implementation. In general, implementation problems are minimized when the project is simple, involving one or two components. There have, however, been several instances when the projects either did not yield full benefits or did not succeed because adequate provision was not made for complementary investments in the scope of the project, e.g., credit and extension in an agricultural project. Because of this experience, during the 1970s, the pendulum swung to the other extreme of including a large number of components in a project. The experience of such projects was also unsatisfactory because of the serious problem of coordination between agencies involved in implementing them. The consensus that has emerged is that a balance should be struck between these two extremes. Even when a multicomponent approach is considered necessary, it may be advisable to concentrate efforts and resources on elements that are essential for achieving the primary objectives, and to leave secondary ones to a subsequent or parallel project.

Project Implementation

The ultimate objective of all preparatory work is to ensure the successful implementation of the project and to realize its intended objectives. If a project is prepared thoroughly, its implementation should proceed smoothly. But events rarely move according to plan. Delays occur in one or more critical areas which slow down the whole process. Or, the objectives themselves may have to be revised in the light of an altered situation. Project implementation, therefore, is the most critical stage in the project cycle. Since project implementation seldom proceeds as planned, sufficient flexibility should be built into a project, so that changes, if found necessary, can be carried out during implementation. The use of sound management techniques such as monitoring of implementation, critical path analysis, and management information system can help in avoiding bottlenecks and delays, thus ensuring the smooth implementation of the project.

The successful implementation of a project greatly depends upon the establishment of a strong and effective project implementation team which can remove bottlenecks and quickly resolve problems which arise during implementation. One of the common problems noted is the lack of coordination among various agencies responsible for implementing the different components of a project. A clear demarcation of responsibility of each participating agency and adequate institutional arrangements and incentives for cooperation are necessary to avoid delays in project implementation.

One of the most frequent problems encountered in project implementation is the shortage of financial resources. This, as noted earlier, is caused by including more projects in a development plan than resources can permit, resulting in all-round underfunding of projects. The problem is compounded by the fact that many public enterprises are not allowed to increase the prices of their goods and services consistent with increased costs of production. The slow release of funds leads to delays in project completion, cost overruns, and reduced contribution to the national output.

Technical problems are another major cause for delays in project implementation or for the nonrealization of the full benefits of a project. Unsuitability of imported equipment for local conditions, design deficiencies, poor quality of materials, and unexpected soil conditions are among the major causes of delay.

In agricultural projects, inadequate appreciation of socioeconomic conditions and cultural practices while introducing new technologies or inadequate technical packages adversely affect the impact of a project. While some of these problems can be mitigated through better preparation, others can be reduced through experience gained in implementing similar projects.

Lack of management and technical expertise is perhaps the most critical factor in slow implementation of projects in many developing countries. This leads to inadequate project supervision, slow decision process, poor coordination, delays in procurement and installation of equipment, etc. There is no easy or quick solution to this problem because it can only be solved through an intensified education and training program in the country. External aid agencies generally provide outside expertise to help implement projects. But even their success greatly depends upon the quality of support provided by local managers and technical personnel.

Post-Evaluation of Projects

The last stage in the project cycle is the evaluation of the impact of the completed project. The purpose is to assess the success or failure of a project through a comparison of the project results with the initial objectives set for it. The post-evaluation report should make a thorough review of the costs and benefits, as anticipated and realized by the project, and explain the reasons for any differences.

The first objective is to benefit from the experience of a completed project so that lessons learned can be incorporated into future projects.

The second objective is to ensure accountability on the part of the project managers. Since it takes several years before the full benefits of a project are realized, at least two post-evaluation studies should be undertaken: one just after project completion and the other four to five years later. The purpose of the first study is essentially to compare the actual cost with the initial cost estimates, while the second study is necessary to compare *ex-post* and *ex-ante* estimate of project viability.

In most cases, the benefits or the impact of the project are measured by the difference in output in the with- and without-project situations. Such a measurement is possible only if an assessment is made of the situation prevailing in the area without the project and then compare it with that which will prevail after the implementation of the project. In the case of an irrigation project, for instance, two separate estimates are necessary. The first estimate should show the existing areas, crops, yields, and inputs, and the expected changes in them as a result of the project. The second estimate should show how the existing area, crops, yields, and inputs will change even without the project. The value of the net increase in output in the first case *minus* the net change in the second case constitutes the benefit of the project. The major problem in estimating the net benefits of a project is estimating the situation that will prevail without the project.

Since it is important to assess the benefits of all projects and their impact on various income groups, it is essential to set up a small benefit monitoring and evaluation unit in each ministry or major department of a government. To ensure objectivity of evaluation, such a unit should report directly to a senior official of the ministry. In smaller countries, such a function could be centralized and made part of the ministry of planning. Since the benefits to be assessed will vary according to the nature of the project, the composition of the unit will vary among ministries. However, it is essential that at least one economist and a statistician be included in each evaluation unit (for further discussion, see chapter 11).

The post-evaluation studies undertaken by international lending institutions for projects financed by them show that the actual impact of projects have, in most cases, been below original expectations. The situation, however, varies considerably among countries and projects. In general, countries with adequate human and financial resources performed better than those where these resources were in short supply. Also, projects in industrial and transport sectors showed better results than those in agriculture and social infrastructure. Some of the major problems generally encountered in implementing projects have been mentioned above. On the benefits side, benefits realized have been low due to a variety of factors. For instance, in the agricultural sector, actual benefits have been lower due to: (1) smaller irrigated areas; (2) slower extension of modern cultural practices; (3) lower yields of crops; and (4) lower prices of products. In evaluating projects, a distinction should be made between factors which are endogenous to the project and which could be improved with better planning and thorough project preparation, and those which are exogenous to the project over which the analyst has no control, e.g., changes in the world price of commodities. The lessons from post-evaluation on account of endogenous and exogenous factors influencing project performance will be different.

CHAPTER 3

Essential Steps in Project Selection

The economic analysis of projects is a method which helps ensure that the development impact of public sector investment projects is maximized. This is achieved by estimating the net contribution of individual projects to national output after which those which will contribute most to national welfare are selected. Before undertaking a detailed economic analysis, it is essential that individual projects are subjected to three major tests.

First, the projects for which detailed feasibility studies are to be carried out should be related to the macroeconomic and sectoral objectives of a country.

Second, the projects selected should adequately meet the demand for specific goods or services in the subsector or area to which they belong.

Third, a comparison should be made among the various mutually exclusive options available for achieving the project objectives and the one which involves the least cost should be selected. These tests are essential for screening because they ensure that those projects selected for detailed appraisal would result in the most efficient use of scarce public funds. The following paragraphs explain briefly the use of these tests in the project selection.

Consistency with Broader Economic Objectives

Economic development is a long-term process and development programs are formulated and implemented at:

- The national level, where national investment plans are formulated, priorities among sectors are established, and macroeconomic policies are put in place.

- The sectoral level, where priorities for investment in each sector are determined and issues and problems affecting the development of such sectors are redressed.
- The project level, where individual projects are identified, prepared, and implemented, with attention given to their technical, financial, institutional, and economic dimensions.

These three levels may also be viewed as three logical stages in the formulation and implementation of public sector development programs. Project work is, however, a continuum and decisions and actions affecting individual projects may take place simultaneously and interactively at each of the three levels. At the national level, changes in the macroeconomic policy framework (fiscal, monetary, exchange rate, wage, and trade policies) can have a major bearing on the economic analysis of individual projects. It is essential that, wherever possible, changes in these policies should be taken into account in assessing the economic viability of projects. On the other hand, project appraisal cannot be directly used for achieving macroeconomic objectives. However, care should be taken to see that the project objectives are not at variance with them.

The relationship between sectoral and project analysis is much more intimate. Sectoral analysis bridges the gap between the macroeconomics of national investment management and the microeconomics of individual projects. It is indispensable for resolving the questions of choice, priority and interrelationship among projects. Some of the policy issues that cut across all projects may be dealt with at the sectoral level. These may include sectoral contribution to overall economic development, the determination of investment priorities in a sector, pricing policies, and institutional matters which have a major bearing on project implementation. At the same time, individual projects can be used as means for achieving sectoral objectives.

Before undertaking an economic analysis of projects, it is essential to define the sectoral objectives and how a certain project intends to achieve such objective. It is essential that the project which can achieve sectoral objectives be given priority in the use of investment resources of the concerned sector. Many bilateral and multilateral institutions have seen several projects falling on the wayside because the policy makers and/or the beneficiaries did not give priority to their output or objectives.

This has resulted in an unnecessary waste of scarce investment resources. The involvement of the beneficiaries is particularly essential in agricultural, rural, and social sector projects.

Particular attention should be given to institutional aspects and pricing policies; care should also be taken to ensure that these are consistent with sectoral objectives. Weak institutional capabilities and a lack of qualified and experienced staff have been major impediments in the implementation and operation of projects in developing countries. Sectoral analysis should help identify these gaps and, wherever necessary, institutional strengthening should be built into the scope of the project. At the same time, inappropriate pricing policies have often resulted in financial insolvency of even economically sound project. Pricing policies of individual projects should be so determined that they ensure financial viability of individual projects and help realize the sectoral goals.

Demand Analysis

Meaning and Method

The next step is to ensure that the size and scale of a project adequately meets the projected needs of the concerned subsector or region. If the scale of a project is too small in relation to the growing demand for its relevant goods or services, scarcities might emerge and adversely affect the sectors for which the project provides the input. On the other hand, if the size of the project is too large in relation to future demand, excess capacity may remain unutilized for a long time, thereby leading to a waste of scarce resources and low financial and economic returns. These economic costs could be avoided through the careful application of demand analysis.

Economists may have two different meanings for the term "demand". The classical one or the schedule of demand concept answers the question of how much of a commodity will be purchased at various prices. Under this concept, demand varies only when the schedule changes. Or that the demand schedule will not change when the quantity that will be purchased or used increases because of a reduction in price. However, the demand schedule will change—or will shift to the right side—if more can be sold at the same price.

The term "demand" may also mean the total quantity purchased or demanded at a given time. In this sense, demand is a function of many variables, in addition to the price of the commodity. These variables may vary depending on the commodities and sectors, but the principal ones include disposable income, population, and prices of competing products.

Economists often interchange two meanings of "demand". As long as both meanings are kept in mind, the double use of the term should not cause any confusion. For instance, in the classical sense, the demand schedule can shift as a result of changes in income, population, and prices of substitutes, while in the total quantity sense, demand is a function of many variables like prices, income, population, and substitute products.

While considering demand in both senses, it is important to make a distinction between the short-term and long-term relationship between demand and the different variables—in particular between price and demand. In the short run, the impact of price on demand is likely to be much smaller than in the long run. For instance, in the power sector, consumers may not shift to alternative sources of energy even if there is a large step-up in prices because of a limited supply of alternative energy sources and of associated equipment required, and a lack of adequate market information. However, with a more flexible supply of energy and related equipment in the longer term, the consumers might shift to alternative energy sources if their total cost is lower than that of electricity.

There are a number of approaches to determine the demand relationship. The most commonly used is the time-series correlation-analysis approach which uses historical data to show how demand or consumption varies with changes in price, income, occupation, and other relevant factors. The simplest technique for approximating the demand curve is to take time-series data showing the level of sales or consumption at various prices over the years and plot the prices as a function of the level of sale or consumption. The price data should be expressed in constant prices of the base year by deflating the current prices of each year by an appropriate price-level index.

Implicit in the above analysis is the assumption that the demand function remains unchanged over time. In actual practice, other factors intervene to change the demand function. In such situations, multiple-regression techniques are used to correlate price and consumption with other variables that measure changes in factors which cause a shift in the demand curve. When multiple factors are involved, variables other than price can be held constant at their average or at any other values to see what the price-consumption relationship would have been if the demand schedule had not shifted over the years. Alternatively, prices can be held constant to see the impact of other factors on the change in demand.

The other approaches used to determine demand relationships are the controlled experiment approach and market research approach. Under the first approach, one variable such as price is changed and all other factors are held constant to study the response of consumers to the price change. Under the second approach, an assessment is made of the response of the consumer based on variables such as price, income group, occupation, and geography. The assessment may be based on personal interviews, written questionnaires, etc. The use of both approaches has been rather limited due either to the high cost of carrying out such studies or questions about the reliability of data.

Demand Analysis for Traded Goods

The importance of demand analysis and its range of factors to consider greatly vary with the different sectors involved. Demand analysis is relatively simple if one talks of goods and services which are bought and sold in the domestic or foreign markets. If the project output is for the international market, an assessment of the price and quality of the product should be made vis-a-vis the competing products and the foreign markets in which it is to be sold. If the output of the project is relatively small in relation to the total traded quantity—as is most often the case—then the world price is the one that is relevant for the project. All that is required is to ensure that the project output can be produced at a price and of a quality that can successfully compete in the world market.

If the project output is for the domestic market, it may be necessary to undertake a market survey to assess total market demand and the existing and potential competitors facing the project. If the project output represents only a small part of the growing market, no detailed demand analysis is necessary, as long as the price of the project output is competitive with that prevailing in the market.

Demand Analysis for Nontraded Goods

Demand analysis is particularly necessary for those sectors and subsectors whose outputs are not traded either in the international or domestic market. These include services provided by the transport sector, power utilities and social sectors. These services are natural monopolies which are controlled by the public sector in most countries. While traded goods can be exported or imported if the demand estimates prove incorrect, the same is not true about the services provided by projects in the economic and social infrastructure sectors. In these sectors, any serious miscalculation about future demand could lead to either excess capacity or shortages with consequent adverse effects on the economy.

In the power sector, the variables most commonly used in estimating future demand are the growth in total domestic output, structure of production, increase in population, power tariffs charged to various consumer groups, and prices of alternative sources of energy. A review of projects approved by international institutions in the past shows that inadequate consideration has been given to the impact of tariffs on power demand and the prices of alternative sources of energy. This has resulted in unrealistic demand estimates for power generation, leading to excess supply when tariffs had to be increased because of high production cost, or shortages when real tariffs were kept low in comparison with the costs of alternative fuels.

The importance of tariffs in determining power demand has, until recently, not been fully recognized not only because of inadequate data but more importantly because tariffs have often not been based on the cost of production. In many countries, power tariffs were kept low in the mistaken belief that these will help accelerate agricultural and industrial production and improve the real income of low-income groups. In actual practice, however, this policy has been proven counter-productive. The losses of power utilities have mounted in many countries, and since governments have been unable to finance these losses and provide resources for future expansion, the demand for electricity has outstripped supply, resulting in load shedding and a general deterioration in the quality of service.

There is a growing recognition that power tariffs should be based on the incremental cost of production, and since these costs are generally rising, tariffs should be raised to cover the rising costs. The increase in tariffs necessary to meet the future demand should be fully taken into account. Furthermore, the prices charged to different consumer groups should be directly related to the cost of supplying power to them. The introduction of economic prices will result in more efficient use of electricity in developing countries. Waste and unnecessary electric use will be minimized, and large consumers will be induced to adopt energy-efficient technologies.

In estimating the demand for electricity, a distinction should be drawn between established and new markets. In the established markets, electricity is already available to consumers, and future demand can be estimated by utilizing information that is available from time-series data. The distinguishing feature of a new market is that, without the project, electricity will not be available to consumers who are using other sources of energy like kerosene for household lighting and diesel sets for commercial or industrial uses. The introduction of electricity in a new market will result in the substitution of other forms of energy, as well as changes in the quality and quantity of energy consumed. Thus, the demand forecast in a new market is likely to be quite different from an established market. Inter-fuel energy norms based on the actual behavior of consumers and increased use due to greater convenience and lower prices should be taken into account in estimating the demand for the new market.

In the transport sector, location-specific variables play a more important role in determining the demand for different transport services. For instance, in a port project, the future demand is a function of growth in output in the hinterland, the importance of international trade to the country concerned, and the nature of import and export products passing through the concerned port. Port charges play a relatively minor role in the estimation of the future demand for port facilities. This is largely because such charges constitute a small fraction of the price of internationally traded products.

Similarly, the demand in road transport projects is highly location-specific. In rural road projects, for instance, the future growth in demand largely depends upon the increase in agricultural production, growth in rural industries, and increase in population in the area serviced by the project. In the case of highway projects, the pace of urbanization, rate of industrial growth, and the increase in rural-urban linkages are important factors in determining future demand. However, in determining the expansion of highway facilities, due account must be taken of the price of alternative modes of transport which may compete with it. For instance, if the railways compete with road transport in the project area, the price of railway services—for goods and passengers—must be considered in determining the future demand for highways.

In the social sectors such as education and health, the future demand for services depend on the policies pursued by the government. If these services are highly subsidized and are brought closer to the people, their demand may increase rapidly. On the other hand, if these services are operated mainly by the private sector and full costs are recovered from the beneficiaries, as may be the case for higher education, the demand will depend upon the level of income of households and future employment opportunities in the country.

Least-Cost Analysis

Having identified the need or demand for a project, the next step is to establish that the proposed project is the least-cost method of attaining the objectives. This means that the net present value of the project benefits must be higher than, or at least equal to, the net present value of mutually exclusive project alternatives. There are usually many project options, which means that if one is chosen, the others have to be dropped. This applies to different designs, technologies, sizes, and time phasing of what is essentially the same project. This also applies to alternative locations and methods of serving the same market.

For instance, a given area could be irrigated through gravity system by building a dam upstream or by having tubewells in the project area. If gravity system is adopted, the tubewells option has to be dropped. All available options of achieving project objectives should be thoroughly examined and it should not be assumed that such options do not exist. The need to compare mutually exclusive options is one of the principal reasons for applying economic analysis from the early stages of the project cycle.

In comparing project options, it is essential that economic analysis be based on economic prices. There is a general tendency for engineers to suggest the most modern production technology that they may be familiar with. It is quite likely that based on prevailing market prices, the proposed technology may yield the lowest cost per unit of production. However, the results may be very different if economic or shadow prices are used.

The above point may be illustrated by using a hypothetical example. Let us assume that country X wants to implement an irrigation project to increase agricultural production. Based on market prices, the engineers might suggest the use of heavy equipment to build the canal system and the rural roads included in the project, while the agronomist may suggest increased production of sugarcane based on prevailing high domestic prices. However, if inexpensive labor exists (based on its opportunity cost), it may prove cheaper and preferable to adopt labor-intensive production technology. Also, based on world prices of sugar and rice, the country might derive greater economic benefit if rice is produced in the project area while increased demand for sugar is met through imports.

Some projects are mutually exclusive in the strict sense that they are alternative ways of producing the same output. For instance, a given amount of power may be produced through the use of hydro or thermal sources or a road may be built connecting two cities following different routes. In such cases, when the project objective is given, it is necessary only to compare the economic costs of available options and select the alternative with the lowest present value of cost when discounted by the opportunity cost of capital. The role of opportunity cost of capital is crucial in determining the least-cost option. The point may be illustrated by taking the following simplified example.

Let us assume that country Y wants to increase its electricity supply to consumers by 50 million kwh to meet the growing demand in the next five years. It has to choose between hydro and thermal power generation to achieve that objective. The hydro project has a higher initial capital cost but lower operating cost. The reverse is the case with the thermal project. Let us further assume that in both projects capital cost is incurred in year zero, and their economic life, during which the operating cost is incurred, is 10 years.

Table 3.1 gives the fixed and operating cost of both projects. The fixed cost of hydro project is ₱1,200 million and its annual operating cost is ₱200 million per year. The fixed cost of the thermal project is only ₱650 million but its annual operating cost is ₱300 million. To arrive at the present value of the total

TABLE 3.1
Choice between Hydro and Thermal Alternative: An Illustrative Example
(Million Pesos)

	Hydro Power Project (A)			*Thermal Power Project (B)*			*Difference (B–A)*		
Year	*Cost Stream*	*Present Value at 10%*	*Present Value at 15%*	*Cost Stream*	*Present Value at 10%*	*Present Value at 15%*	*Cost Stream*	*Present Value at 10%*	*Present Value at 15%*
0	1,200	1,200.0	1,200.0	650	650.0	650.0	–550	–550.0	–550.0
1	200	181.8	174.0	300	272.7	261.0	100	90.9	87.0
2	200	165.2	151.2	300	247.8	226.8	100	82.6	75.6
3	200	150.2	131.6	300	225.3	197.4	100	75.1	65.8
4	200	136.6	114.4	300	204.9	171.6	100	68.3	57.2
5	200	124.2	99.4	300	186.3	149.1	100	62.1	49.7
6	200	112.8	86.4	300	169.2	129.6	100	56.4	43.2
7	200	102.6	75.2	300	153.9	112.8	100	51.3	37.6
8	200	93.4	65.4	300	140.1	98.1	100	46.7	32.7
9	200	84.8	56.8	300	127.2	85.2	100	42.4	28.4
10	200	77.2	49.4	300	115.8	74.1	100	38.6	24.7
Total	3,200	2,428.8	2,203.8	3,650	2,493.2	2,155.7	450	64.4	–48.1

Equalizing Discount Rate = 10 + 5 {64.4/[64.4 – (–48.1)]} = 12.9

cost (capital and operating) incurred during the life of both projects, the discount rate, which is a measure of the opportunity cost of capital, should be determined. As shown in table 3.1, if the discount rate is 10 per cent, then the hydro project is preferable because at that rate the present value of its total cost is lower than that of thermal project. However, if the discount rate is 15 per cent, then the thermal project becomes a preferred option.

In practice, the least-cost analysis is expressed in terms of the equalizing discount rate (EDR) or the cross-over discount rate. The EDR is the rate at which the present value of two cost streams is equal. In the abovementioned example, the EDR is 12.9 per cent. If the opportunity cost of capital is below this rate, then the hydro project, whose present value of cost would be lower, should be selected. However, if the opportunity cost of capital is above 12.9 per cent, then the thermal option should selected.

It should be noted, however, that the mere fact that a project has the least cost is no evidence that a project is economically viable or that the benefits that will be derived from its implementation will exceed the cost. Therefore, after the least-cost option has been identified, the selected project must be tested for its economic viability and should be implemented only if its benefits exceed cost.

In comparing project options, care should be taken to ensure that the options considered are realistic and strictly comparable. For instance, it would be incorrect to compare two road project options wherein the first involves the use of gravel surface and has an economic life of only five years and the second involves cement concrete and has an economic life of 30 years. If the time dimension is ignored, the gravel road option may appear preferable, while if the additional investment required to maintain the gravel road in good condition for 30 years is taken into account, the second option may appear superior. Also, as mentioned earlier, economic prices should be consistently used in making the comparison. Using economic prices in some cases and financial in others will lead to misleading and incorrect results.

CHAPTER 4

Identification of Costs and Benefits

Having briefly discussed the stages in the project cycle and the essential steps in profit selection, the methodology of economic analysis of projects will now be examined. This analysis basically involves three steps: (1) identification, (2) valuation, and (3) comparison of a project's costs and benefits to determine its viability. This chapter will deal with the first step and will attempt to explain the method of identifying and measuring the benefits and costs of a project.

A useful starting point in this context is to estimate the flow of the financial costs and benefits of a project from its financial accounts. The next step is to indicate various adjustments that should be made to arrive at the economic cash flow of the project. Let us begin with the derivation of a financial cash flow of a hypothetical project, and then examine the type of additional data and adjustments that are necessary to arrive at the project's economic cost and benefit flows.

Estimate of Financial Cash Flow

In assessing the financial position of a project, a financial analysis will have to be done. It relies on the following: (1) the Balance Sheet; (2) the Income and Expenditure Statement; and (3) the Sources and Application of Funds.

The balance sheet is a statement of the assets and liabilities position at the end of a financial year. The income and expenditure statement (hereafter called the income statement) summarizes the current receipts and expenditures during the financial year. The sources and uses of funds statement shows how assets (including operating profits) are used to finance liabilities (including debt servicing).

The primary concern of a project entity, like most enterprises, is to ensure that its annual income covers all its expenses including interest payments and tax liabilities, and leaves adequate profit in relation to its investment. The financial statements are, therefore, drawn from the point of view of the owners of the project.

The main principles involved may be explained by the following illustrative example. Let us assume that there is a chemical plant which is built in 2 years and has an economic life of 5 years, after which its scrap value is nil. Total cost of this project, including an inventory requirement of ₱1 million, is ₱15 million. Of the total cost, ₱6 million is financed from owner's equity, ₱8 million from long-term borrowings, and ₱1 million through short-term liabilities. The balance sheet of the plant at the end of the first and second year would look somewhat like what is shown in table 4.1.

The summary information in table 4.1 is considered adequate for purposes of financial accounting. However, more details are required, especially under buildings and equipment, for the identification and subsequent valuation of the economic cost

TABLE 4.1
Balance Sheet of the Chemical Plant
(Year-End Data in Thousand Pesos)

Liabilities	*First Year*	*Second Year*	*Assets*	*First Year*	*Second Year*
Equity	4,000	6,000	Fixed Assets	9,000	14,000
Capital	4,000	6,000	Land	1,000	1,000
Retained Earning	–	–	Buildings and Equipment	8,000	13,000
Long-term Debt	5,000	8,000	Depreciation	–	–
Debentures	2,000	3,000			
Mortgage Debt	3,000	5,000	Current Assets		
			Inventories	–	1,000
			Account Receivables	–	–
Current Liabilities					
Bank Loans	–	1,000	Cash	–	–
Accounts Payable	–	–			
Tax Due	–	–			
Total	9,000	15,000	Total	9,000	15,000

of the project. A minimum breakdown on the lines given in table 4.2 is essential for economic analysis and should be secured when the feasibility report of the project is prepared. Essentially, the cost of physical investments in construction and machinery should be broken down into imported and local cost components, while the tax component of imported and locally procured goods should be shown separately. The local cost of buildings and installations should be further divided into domestic and imported materials and wages of skilled and unskilled labor.

Let us assume that the chemical plant starts operation on the beginning of the third year and completes its economic life at the end of the seventh year when it has to be scrapped, leaving no residual value. A hypothetical income statement of the chemical plant is shown in table 4.3. According to this statement, the total output of the plant is estimated at ₱8 million in year 3, which will go up to a maximum of ₱14 million in the fifth year, and remains at that level in the sixth year, but declines to

TABLE 4.2
Breakdown of Investment Cost of Building and Equipment
(Year-End Data in Thousand Pesos)

Item	*Year 1*	*Year 2*
Building and Equipment Installation	3,000	5,000
Imported Materials (net of taxes)	1,500	2,500
Import Duties	500	800
Sales Taxes on Imports	100	170
Domestic Materials (nontraded)	480	830
Taxes on Domestic Materials	20	35
Skilled Labor	100	175
Unskilled Labor	300	490
Imported Equipment	5,000	8,000
FOB Value of Imported Equipment	4,000	6,400
Import Duties	800	1,280
Sales Taxes	200	320
Total Building and Equipment	8,000	13,000

₱13 million in the last year. Details on value of goods produced, expenses, operating profits, and their allocation are shown in table 4.3.

To estimate the economic value of the output and of various inputs required to produce that output, it is necessary to have a further breakdown of some of the items mentioned in table 4.3. The items for which further details are essential are the value of sales, material inputs, and labor. A minimum breakdown of information necessary under these headings for economic analysis is shown in table 4.4.

TABLE 4.3
Income Statement of the Chemical Plant
(Thousand Pesos)

Item	*Year 3*	*Year 4*	*Year 5*	*Year 6*	*Year 7*
I. Value of Goods Sold	8,000	12,000	14,000	14,000	13,000
II. Material Inputs	2,000	3,000	3,500	3,500	3,200
III. Operating Expenses	1,800	1,950	2,100	2,100	1,950
Electricity	400	450	500	500	450
Other Utilities	80	100	100	100	100
Services	200	200	200	200	200
Labor	720	800	900	900	800
Repairs and Maintenance	400	400	400	400	400
IV. Depreciation	2,000	2,000	2,000	2,000	2,000
V. Operating Profit (I–II–III–IV)	2,200	5,050	6,400	6,400	5,850
VI. Interest Payments	600	600	600	600	600
VII. Net Operating Profit before Tax (V–VI)	1,600	4,450	5,800	5,800	5,250
VIII. Company Tax	200	1,050	1,300	1,300	1,250
IX. Profit after Tax	1,400	3,400	4,500	4,500	4,000
Dividend	400	1,400	1,800	1,800	1,600
Retained Profit	1,000	2,000	2,700	2,700	2,400

TABLE 4.4
Breakdown of Selected Items in Income Statement
(Thousand Pesos)

Item	*Year 1*	*Year 2*	*Year 3*	*Year 4*	*Year 5*
Total Value of Sales	8,000	12,000	14,000	14,000	13,000
Unit Value (₱/kilo)	10	10	10	10	10
Quantity (kilos)	800	1,200	1,400	1,400	1,300
Material Inputs	2,000	3,000	3,500	3,500	3,200
Imported	1,000	1,500	1,750	1,750	1,600
Local	700	1,050	1,200	1,200	1,150
Duties on Imports	200	300	350	350	300
Taxes on Local Inputs	100	150	200	200	150
Labor Cost	720	800	900	900	800
Skilled	250	300	350	350	300
Unskilled	470	500	550	550	500

From tables 4.1 and 4.3, it is possible to prepare the cash flow statement from the viewpoint of the firm. In preparing this statement, two points deserve to be noted. First, all payments by the project entity are outflows and all receipts are inflows. An increase in accounts receivable represents cash outflow, while an increase in accounts payable represents cash inflow. Second, depreciation is treated as an item in the income statement. The advantage of doing this is that it allows the firm to set aside funds for the eventual replacement of fixed assets and it reduces the tax liability because firms are normally taxed on profits net of depreciation which is treated as an item of expenditure. However, depreciation does not constitute a cash outflow and is, therefore, excluded from the financial cash flow profile of the project. Also, since the full capital cost of the project is already included in the cash flow profile of the project, any further depreciation charge would constitute double counting of the project cost.

The cash flow statement from the viewpoint of the firm is given in table 4.5. A distinction has been made between operating flows and nonoperating flows. The items included in the oper-

TABLE 4.5
Financial Cash Flow of the Chemical Plant
(Thousand Pesos)

Item	*Year 1*	*Year 2*	*Year 3*	*Year 4*	*Year 5*	*Year 6*	*Year 7*
Operating Flows	-9,000	-6,000	1,200	3,050	3,700	3,700	3,450
Sources	0	0	4,200	7,050	8,400	8,400	7,850
Operating Profit	0	0	2,200	5,050	6,400	6,400	5,850
Depreciation	0	0	2,000	2,000	2,000	2,000	2,000
Application	9,000	6,000	3,000	4,000	4,700	4,700	4,400
Net Current Assets	0	1,000	3,000	4,000	4,700	4,700	4,400
Fixed Assets	9,000	5,000	0	0	0	0	0
Non-Operating Flows	9,000	6,000	-1,200	-3,050	-3,700	-3,700	-3,450
Sources	9,000	6,000	0	0	0	0	0
Equity	4,000	2,000	0	0	0	0	0
Borrowing	5,000	4,000	0	0	0	0	0
Application	0	0	1,200	3,050	3,700	3,700	3,450
Interest Payment	0	0	600	600	600	600	600
Loans Repayment	0	0	0	0	0	0	0
Taxes	0	0	200	1,050	1,300	1,300	1,250
Dividends	0	0	400	1,400	1,800	1,800	1,600

ating flows result from the productive operations of the project. It is from these operations that the pre-tax financial NPV and the FIRR are calculated. It is, however, possible to calculate the pre-tax and post-tax return on equity. The former can be calculated by deducting both the borrowed fund and interest payments on them from the operating flows. The latter can be calculated by making a further deduction of the company tax from net operating flows as shown in table 4.5.

Economic Costs and Benefits: General Issues

In preparing the economic cash flow estimate, several adjustments need to be made to the financial cash flow data given in table 4.5. Certain adjustments apply to both the cost and benefit streams of projects, while others are specific to either the cost or the benefit side. In this section the general adjustments are discussed, while adjustments specific to either the cost or benefit side are dealt with in subsequent sections.

The adjustment can be meaningfully carried out only if we fully understand the real economic impact that a project makes on the economy. When a project is implemented, it reduces the supply of inputs (consumed by the project) and increases the supply of output (produced by the project) and thereby changes the availability of inputs and outputs in the economy compared with what it would have been without the project. Identifying the difference between the availability of inputs and outputs with and without the project is the basic method of identifying the costs and benefits of the project.

"With" and "Without" Comparison

The comparison of the "with-" and "without-" project situations should be distinguished from the "before-" and "after-" project situations. For instance, certain changes may take place in both the inputs and outputs even if the project is not implemented. These changes, to the extent that they can be anticipated, should not be attributed to the project and should be excluded from the estimate of its costs and benefits. For instance, in a water supply project, part of the increase in output may be the result of investments already made. In estimating the benefits of the project, the output attributable to previous investments should be excluded, since this output will, in any case, take place even if the project is not implemented. Similarly, if agricultural production is expected to increase without the project due to better seed varieties, improved extension services, or fertilizers, the increase should be excluded from the benefits attributed to the project. The "before" and "after" comparison fails to take into account these changes.

Transfer Payments

Economic analysis excludes from both the cost and benefit streams all transfer payments, i.e., payments made by one sector in the economy to another, because these transfers do not impose any direct claim on the country's resources. Taxes paid by the project entity on the import of machinery or subsidies given by the government to the farmers for food production are examples of such transfers. Interests paid by the project entity on domestic borrowing are considered as transfer payments as they only involve a transfer of purchasing power from the project authority to the lender of capital. Similarly, borrowing or repayment of loans is merely financial transfer and does not impose any economic costs on the country. Royalty payments within a country also constitute transfer payments, but if these payments are made abroad in foreign exchange, they should be treated as an economic cost since the foreign exchange used to make the payments will no longer be available for use by the others.

Externalities

The implementation of a project may, in certain cases, lead to positive or negative effects on the economy and these are not reflected in the financial accounts of the project. If these effects, known as "externalities", involve significant economic costs or confer significant economic benefits, they should be taken into account in estimating the overall net economic impact of a project.

On the cost side, the external economic impact may, for instance, include increased pollution caused by a chemical or cement plant, or the adverse effects of chemical fertilizers on health or fisheries. External benefits may include health benefits of a water supply scheme or recreational benefits of a storage dam. Wherever possible, attempts should be made to internalize the externalities by considering a package of closely related activities as one project. Even where it is not possible to fully measure the externalities, an attempt should be made to identify them and, if possible, to quantify and evaluate them.

Identification of Economic Costs

In the case of an individual firm, project costs are defined to include all payments that reduce the cash receipts or contribute to the cash outflow of a project. Under this definition, all cash payments made for machinery, equipment, materials, rent, energy, labor, patents and royalty fees, taxes and import duties, etc., constitute the cost of the project. In economic analysis, costs include only those items which, when used, affect the availability of resources to the rest of the economy. Under this definition, transfer payments—or payments matched by corresponding receipts by the rest of the economy—are not treated as costs. Similarly, if employment of unskilled labor does not reduce its availability to the rest of the economy due to large-scale unemployment, the economic cost of labor for the project is zero. However, if the project imposes any real cost in the nature of externalities, these should be treated as a cost of the project.

In estimating the economic cost of a project, it is essential to identify all the relevant inputs of goods and services, both by way of fixed investment and of operating costs required to achieve the benefits of the project. When these inputs are used for the project, they correspondingly reduce their availability to the rest of the economy. The use of inputs by the project could result in either one or a combination of the following two effects. First, the use of certain inputs may result in a decline in their availability exactly equal to their consumption by the project. This is rather an exceptional situation and would arise only when the national availability is fixed either because of an acute foreign exchange situation or because the prices of those products are fixed by the government below the cost of their production, making additional output nonviable. Second, in response to the input demand for the project, the supply may correspondingly increase, causing no effect on the rest of the economy. These two types of inputs should be separately identified because the method of valuing them is different.

There are three cost items which merit special attention. These are "contingencies", "sunk cost", and "depletion premium". They are briefly discussed below.

Contingencies

In estimating the financial cost of a project, provision is often made for physical and financial contingencies to allow for (1) additional inputs that may be required to complete the project and for (2) additional financial costs that may result from price increases during project implementation. While contingency allowances are determined by engineering and financial considerations, they have an implication for economic analysis. Since the economic rate of return is measured in constant prices, provision for general price increases should be excluded from the economic cost of the project. However, if allowances have been made for changes in relative prices, these should be taken into account since such changes alter their claim on the real resources of a country. Physical contingencies should be included in the economic cost of a project because they represent the monetary value of additional real resources that may be required beyond the base cost needed to complete the project.

Sunk Cost

A project may require the use of facilities or assets in existence prior to the appraisal of the project. The cost of such facilities is a "sunk cost" and should not be included in the project cost, neither should it be used to determine whether or not to proceed with a project. It is assumed that the existing facilities have no alternative use and, therefore, do not involve any opportunity cost. However, previous investments should be considered in the appraisal of a new project if the existing facility would close down in the absence of new investment. In such a case, it is not the historical cost which is relevant but the liquidation value of the existing assets which should be included in the cost of the new project. Correspondingly, the benefits of the existing investment in this case should be included as benefits from the project.

In some cases, a project constitutes part of a sequence of related investments. While a project that uses excess capacity created by an earlier project may well show high returns, such returns may also arise if a project is designed in a way that allows it to capture benefits originally expected from an earlier project. For instance, a rehabilitation and modernization project of an irrigation system may include as benefits yield increases expected from the original project. Hence, in such cases it is desirable also to indicate the net return on the entire project, including sunk costs, in order to show whether the original decision to provide the facilities was well founded.

Depletion Premium

Some projects involve the exploitation of nonrenewable natural resources such as crude oil, natural gas, and mineral deposits. The opportunity cost of using such natural resources must be taken into account in the economic analysis of such projects. Such resources cannot be replenished and, when depleted, must be replaced by imports or domestic substitutes. The opportunity cost of using such a resource is, therefore, the cost of its substitute when the resource is exhausted. The depletion premium or allowance is the economic price of using such a resource and depends upon the proportion of total reserves exploited during each year.

Identification of Economic Benefits

In financial analysis, all additional income resulting from the operation of a project is treated as benefits. In economic analysis, the identification and measurement of economic benefits are somewhat more complex. The nature of the benefits varies with projects. In some projects, the benefits are measured by the net increase in output contributed by the project. In others, the benefits may be measured by the saving in costs which would otherwise have been incurred in the without-project situation. In certain cases, the benefits may be represented by the supply of goods and services for which there is no freely operating market or whose value cannot be expressed in monetary terms. The benefits should be carefully identified and measured because the valuation method varies with the nature of project benefits.

Consumer Surplus

In some cases, the project output may be large enough to reduce the price of the output paid by consumers compared with the level that would prevail without the project. The reduced price used in project accounts understates the economic benefits of the project since the savings of the consumers on intramarginal units will be ignored. The difference between what the consumers are willing to pay and what they actually pay is called "consumer surplus". This surplus accrues to the consumers when a project contributes to a reduction in price and should be treated as a project benefit. In cases where it is not possible to quantify this surplus, a qualitative assessment of this benefit may be provided.

System Analysis

If a project is part of a larger system, the full benefit of the project may not be realized unless matching investments are made in other parts of the system. For instance, benefits from a power generation project may not fully materialize unless certain investments are made in transmission and distribution. Similarly, a road project may need complementary investment in bridge construction to achieve the full benefits of the project. In financial analysis, it may be assumed that related investment will be made by another entity. In economic analysis, however, the boundary of the project should be redrawn and all investments necessary to achieve the objectives of the project should be considered as a package and should be included in the project analysis.

Classification of Benefits

Based on the criteria of measurability, public sector projects may be classified into four broad groups: (1) agricultural and industrial projects, (2) transport projects, (3) public utility projects, and (4) social projects. The identification and measurement of benefits in the agricultural and industrial sectors are the easiest. Here, the benefits are measured by the increase in output contributed by the project.

For instance, in an agricultural project, the benefits are measured by the increase in output of various crops over the level that would prevail without the project (see example in chapter 13). In a new industrial project, the entire output represents the benefit of the project.

Identification of benefits is relatively more difficult in transportation projects. In highway projects, a distinction should be drawn between the benefits accruing due to the traffic that would exist without the project and that which will be generated as a result of the project. In the former case, the major benefit would be the savings in the operating cost of the vehicles due to reduced wear and tear, savings in fuel consumption, lower maintenance cost, and longer life of the vehicles. To this should be added the value of working time saved and the reduction in road accidents.

It is, however, more difficult to assess the new traffic generated as a result of the project and apportion that benefit to the road project. For instance, a rural road project may lead to an increase in agricultural output due to better access to markets, but it is very difficult to determine the increase in traffic flow and estimate the corresponding benefits attributable to the project. Road transport project benefits thus consist of two types: one is existing demand that can be identified and measured accurately and the other is new demand which is more difficult to predict and involves a measure of judgment.

Similar difficulties arise in identifying and measuring benefits of a port or an airport project. For instance, in a port project, a reduction in ship turn-around time serves as a partial measure of the benefits of that project. But complex issues of measurement arise where the major effect is the generation of additional economic activity in the hinterland. Here, an assessment has to be made of the portion of economic benefits which can be attributed to the establishment of the port facility. Special problems arise where the benefits do not accrue directly to the country in which the investment is made but to the foreign shipping lines. In such cases, a judgment has to be made based on market conditions on the extent to which the benefits of the project will flow back to the country through a reduction in shipping cost.

Public utility projects raise more complex issues of benefits measurement. For instance, the quantity and price of water supplied do not serve as a useful measure of the benefit of a water supply project. A rural water supply project may improve the health of the population by reducing the incidence of waterborne diseases caused by untreated water. The project may also greatly reduce the time spent in bringing the water from the village well or from the river stream. Such benefits cannot be easily measured in quantitative terms. A judgment has to be made regarding the value placed on health and the time spent on fetching water to arrive at the estimate of the full benefit of a rural water supply project. Similarly, in the case of a rural electricity scheme, the quantity and price of electricity consumed do not yield a good measure of the benefit of the project. The correct measure is provided by the reduction in the requirement of alternative sources of energy such as kerosene, oil, wood, and diesel oil—whose use is replaced by electricity.

Social sector projects, including health and population, education, sewerage, and poverty reduction programs, pose the most complex issues of measurement. For instance, many education projects are intended to improve the quality of education provided in existing institutions. However, even with large-scale investments in teachers' training, availability of better equipment, and library facilities, it is very difficult to measure the actual improvement brought about by the project. Similarly, in a health project, the primary benefits are measured by the reduction in the morbidity and mortality of the population in the project area. But it is very difficult to ascribe the benefits to the project and, more importantly, to translate them into monetary terms. It is for these reasons that a full-blown cost-benefit analysis for social sector projects is normally not carried out and, instead, a least-cost or cost-effectiveness approach is adopted (for further discussion, see Cost-Effectiveness Analysis later in this chapter).

Presentation of Economic Cash Flow

Having discussed the difference between what constitutes the financial and economic costs and benefits of a project, an attempt is now made to prepare the economic cash flow statement based on the data given in tables 4.1 to 4.5. It may be noted that the nonoperating cash flows are not relevant in economic analysis. This is because in public sector projects with which we are primarily concerned, a distinction between equity and loan funds is not particularly relevant as all funds are provided by the government. Consequently, separate figures for interest and dividends are not shown. Also, as was already explained, all transfers are excluded from the economic cash flow. Therefore, only operating flows which include the value of output and the operating cost of the project are shown in the economic cash flow statement of the project.

In economic analysis, we are much more concerned with the components of the cash flow statement. The purpose is to identify items which should either be excluded from or added to the financial cash flow statement or which require reevaluation compared with the data given in table 4.5. For reasons already noted, taxes and duties on investment goods, on inputs, and on outputs should be excluded for estimating the economic cash flow. The cost of land used for setting up the chemical plant is also not treated as an economic cost if its use does not involve any reduction in real resources to the rest of the economy. This will be the case when the land is barren and cannot produce any agricultural or mineral output which could be considered as its opportunity cost.

Financial analysis does not provide for the treatment of affluents which imposes a real cost on the rest of the economy.

The financial cost of the outlay required for the installation of an affluent treatment unit is estimated to cost ₱1 million. Making the abovementioned adjustments and adding the breakdown of the items mentioned in tables 4.2 and 4.4 allows one to arrive at the economic cash flow statement of the chemical plant, as shown in table 4.6.

TABLE 4.6
Economic Cash Flow of the Chemical Plant
(Thousand Pesos)

Item	*Year* 1	*Year* 2	*Year* 3	*Year* 4	*Year* 5	*Year* 6	*Year* 7
Building and Equipment Installation	–1,980	–1,350					
Imported Materials	–1,500	–1,000					
Local Materials	–480	–350					
Skilled Labor	–100	–75					
Unskilled Labor	–300	–190					
Imported Equipment Including Treatment Unit	–4,000	–3,400					
Inventories	0	–1,000					
Imported Material Inputs			–1,000	–1,500	–1,750	–1,750	–1,600
Local Material Inputs			–700	–1,050	–1,200	–1,200	–1,150
Electricity			–400	–450	–500	–500	–450
Other Utilities			–80	–100	–100	–100	–100
Services			–200	–200	–200	–200	–200
Skilled Labor			–250	–300	–350	–350	–300
Unskilled Labor			–470	–500	–550	–550	–500
Repairs & Maintenance			–400	–400	–400	–400	–400
Total Cash Outflow	–6,380	–6,015	–3,500	–4,500	–5,050	–5,050	–4,700
Total Cash Inflow (Sales)			8,000	12,000	14,000	14,000	13,000
Net Cash Flow	–6,380	–6,015	4,500	7,500	8,950	8,950	8300

Cost-Effectiveness Analysis

The previous sections explained the method of identifying the items of economic costs and benefits of projects. The subsequent steps in valuing benefits and costs in economic prices and determining the economic viability of the project are discussed in the following chapters. Such an analysis is, however, meaningful only with respect to projects for which both the costs and benefits can be fully or largely identified and valued in monetary terms. There are, however, several projects where it is not possible to provide a monetary measure for most of the benefits realized. This is particularly true of most projects in the social sectors like education, health, and water supply. In determining the choice of projects in these sectors, cost-effectiveness analysis is generally used. Cost-effectiveness analysis takes account of both the costs and effects of the projects and helps achieve one of two objectives: (1) given the end results, it helps select the least-cost method for achieving such results and (2) given a fixed amount of available resources, it helps maximize the end results.

Under cost-effectiveness analysis, both the costs and effects of different alternatives are evaluated to determine the choice of a project. This analysis can be applied only where (1) programs have identical or similar goals which can be compared and (2) a common measure of effectiveness can be used to assess them. The cost data can be combined with the effectiveness data to provide a cost-effectiveness evaluation that will enable the selection of that alternative which will yield the maximum effectiveness per unit of cost, or which requires the least cost per level of effectiveness.

A simple example is given to illustrate the use of cost-effectiveness analysis. Let us assume that the objective is to improve the science education scores of low achievers in high schools. Five schools are chosen in a country, and 100 students are randomly assigned in each school in groups of 25 students each to three remedial instructional treatments, while one control group of 25 receives no remedial instruction. The alternative instructional treatments selected for comparison are: (1) a remedial group working with a special instructor (GAP); (2) an individually programmed instructional curriculum (IPI) in which each student works at his own pace in a special room with individualized material and with a coordinator; and (3) a peer-tutoring (PT) approach in which senior and bright students spend an hour a day with a small group of five students each.

The monthly costs of the various instructional treatments are assessed by determining the various ingredients used and their market value. These are shown in table 4.7. The cost per student in the first approach is ₱500, while that of the second approach is ₱900. The third approach involving peer tutoring has the lowest cost of ₱100 per student.

TABLE 4.7
Cost-Effectiveness of Remedial Science Education Programs

Method	*Cost per Student*	*Effectiveness (Test Score)*	*C/E*
Group Approach Program	500	5	100
Individually Programmed Instructions	900	20	45
Peer-Tutoring	100	2	50

At the end of the year, the students in all the five schools are tested to evaluate the effectiveness of each program. Effectiveness is measured by the difference between the average of their scores and those in the control group. The hypothetical examples in table 4.7 show an effectiveness of 5 points under GAP, 20 points under IPI, and 2 points under PT. When these results are combined with those of costs, a cost-effectiveness ratio is obtained. The ratio shows the cost per student of an average one point improvement in test scores. The cost-effectiveness results show that based on the cost factor alone, the PT approach appears to be preferable since its cost per student is the lowest among the three alternatives. However, when effectiveness is taken into account, the IPI program turns out to be the most preferred approach because the cost per student for improving their score—the objective of the project—is the lowest under this alternative.

It should be noted that cost-effectiveness analysis is useful only when cost analysis can be combined with data on effectiveness. Where the effectiveness is either not specified or cannot be measured, cost-effectiveness analysis is reduced to least-cost analysis. For instance, if the objective of a project is to provide potable water to rural communities, without considering the security of supply, only least-cost analysis can be undertaken among possible alternatives.

In conclusion, there are certain limitations in the cost-effectiveness analysis. First, one can compare C/E ratios among alternatives with only one goal. It is not possible to compare alternatives with different goals like history versus science or education versus health projects. Second, C/E analysis, by itself, says nothing about the efficiency of the resource use. Based on such analysis, it is not possible to decide whether, for instance, the scarce resources should be used for improving science scores or for increasing the literacy level in the country. The issue of priority among alternative objectives can be decided only in the context of sectoral or community needs.

CHAPTER 5

Economic Valuation of Goods and Services

Introduction

After identifying the costs and benefits and establishing the size of their flows over the life of a project, values should be attached to them so that they can be aggregated and compared. The costs and benefits should be valued according to their economic prices, which, in most cases, differ from their market prices. The process of converting market prices to economic or accounting prices is the most important and also the most difficult part of the economic evaluation of projects.

Since the main objective of economic analysis is to assess the real contribution that a particular project is expected to make to the national economy, costs and benefits should be valued in constant prices, i.e., in terms of the prices prevailing in the year in which the project is appraised. Any expected changes in the general price level during the life of the project should be disregarded, but anticipated changes in relative prices should, as explained in chapter 7, be taken into consideration. The economic prices of inputs used and output produced by a project depend on the value that a society places on the inputs that are withdrawn from and on the outputs that are added to the economy. It should be noted that these prices relate to an economic environment where distortions are expected to persist, not the equilibrium prices that would prevail in the absence of distortions. This does not necessarily imply a passive acceptance of the existing distortions. Any expected policy changes which would reduce or increase these distortions should be taken into account.

This and the next three chapters will discuss the method of valuing the economic prices of the various project inputs and outputs. The ultimate purpose of the entire exercise is to convert all values into border prices. This is done on the assumption that international trading opportunities provide the basis for calculating the economic worth of both the domestic output and the factors of production in the national economy. Using international trade possibilities to value all domestic inputs and outputs is not based on the concept of free trade, nor does it suggest that prices of internationally traded goods are free from distortions. Rather, it reflects the viewpoint that foreign trade provides a country with the real opportunity to sell or purchase merchandise and that these opportunities should be recognized in public investment policy.

Economic Prices of Traded Goods and Services

It is necessary to start by making a distinction between goods and services that are traded internationally at the margin and those that are not. These will be referred to as "traded" and "nontraded" goods and services, respectively. Traded goods are goods and services that are actually imported to or exported from the country. They are not subject to binding quantitative restrictions such as import quotas or prohibitive taxes (i.e., taxes which are so high as to prevent trade from occurring). All the other goods and services are nontraded. The valuation for these two categories is different.

To decide whether or not a good is traded, one should look at its ultimate impact on exports or imports. When additional demand for a good is met entirely by importing the product or diverting it from exports, that good is directly traded. A good is considered traded even if it is produced domestically but part of the existing demand is met through imports. Similarly, a good is considered traded if a country is a net exporter of a good and any additional output from a project will further add to existing exports.

The determination of the correct economic price for traded goods is important. It depends upon (1) whether the good is imported or exported, (2) the point of marketing within the country, and (3) whether the quantity sold or purchased affects its price. In most cases, purchasers' price at the point of delivery which includes trade and transport marginsis used. Generally, the amount of a commodity used as an input or produced as an output by a project constitutes only a small fraction of the total supply available in the world market and has, therefore, no effect on world prices. This assumption could be made for most goods unless there is evidence to the contrary.

Economic Price of Imported Goods

If a project uses an imported input, or produces an output which replaces part of an imported commodity, the economic or "accounting" price will be based on the landed cost, or import cost plus insurance and freight (CIF). Transport and distribution charges should be added to the CIF price, expressed in the local currency equivalent at the official exchange rate. This is done to bring the commodity from the port of entry to the point of consumption, either by the project entity or the households. Since transport and distribution costs are expressed in the local currency, they must be converted into their border price equivalent before they are added to the CIF price. This is done through the use of appropriate conversion factors, a subject discussed in detail in chapter 8. The rule for determining the economic price of imported good at purchasers' price is thus:

(Economic price of imported goods) = CIF + (Transport and handling costs expressed in border price equivalent) (1)

The procedure for determining the economic price for an imported good is illustrated in table 5.1. The column headings of the table are self-explanatory, except for the column conversion factor, which converts the domestic market price of a nontraded good into its border price equivalent. The CIF import price of the good is estimated at $150 per ton, or ₱3,000 in local currency.

Examining the individual components of the cost of an imported good provides the reasons for the difference between its market and economic values. The economic prices reflect the real cost or value of resources to the national economy: import tariffs do not constitute an economic cost. The economic costs of transport and distribution are also estimated to be lower than their corresponding domestic currency values, because of the high import duties which greatly increase the domestic prices of vehicles and fuel. On the basis of the calculations in table 5.1, the economic price in local currency is estimated at ₱3,440, while the market price works out to be ₱4,150. Most of the difference between market and economic value is due to the import tariff which, while included in the market value, is excluded from the economic value.

TABLE 5.1
Market and Economic Value of an Imported Good

Item	*Dollars ($)*	*Exchange Rate*	*Domestic Currency (Pesos)*	*Market Value (Pesos)*	*Conversion Factor*	*Economic Value (Pesos)*
CIF Price	150	₱20 = $1	3,000	3,000	1	3,000
Tariff	–		500	500	0	–
Transport	–		400	400	0.6	240
Distribution	–		250	250	0.8	200
Total				4,150		3,440

Economic Price of Exported Goods

If a project uses as an input a good which is exported or produces a good which adds to exports, then the economic price will be based on the free on board (FOB) export price less local transport and handling charges expressed in border price equivalent. The rule for determining the economic price of exported output is the following:

(Economic price of exported output) = FOB – (Transport and handling costs expressed in border price equivalent) (2)

The economic price of a good diverted from export (to domestic market) for consumption or for use as an input involves the determination of two things: the cost saved and revenue foregone by not exporting, and the transport and handling costs incurred in using the good in the domestic economy. This is derived by adding to equation 2 above the transport and handling cost to the point of domestic delivery, expressed in border prices. Table 5.2 explains the method involved in calculating the economic price of exported goods under both situations.

Based on the data given in table 5.2, the economic value of an exported good based on purchasers' price is ₱970. This is derived by deducting the ₱30 export handling charges from the FOB export price of ₱1,000. The economic price of the good diverted from export for domestic consumption is derived by adding to ₱970 the economic value of the transport and distribution charges on domestic sales, estimated at ₱39. This works out to be ₱1,009.

Traded Goods with Variable Prices

The procedure for determining the economic prices of traded goods as described above assumes that world prices remain unaffected by the traded inputs or output of a project. Such an assumption is valid in most cases because, in general, the im-

TABLE 5.2
Market and Economic Value of an Exported Good

Item	*Dollars ($)*	*Exchange Rate*	*Domestic Currency (Pesos)*	*Market Value (Pesos)*	*Conversion Factor*	*Economic Value*
FOB	50	₱20 = $1	1,000	1,000	1	1,000
Export Tax			200	200	0	0
Transport for Export 30				50	50	0.6
Ex-Factory Price			750	750	1	970
Transport for Domestic Sale			25	25	0.6	15
Distribution for Domestic Sale			30	30	0.8	24
Total				805		1,009

pact of individual projects on world trade is very small. This position may, however, not hold when a country imports or exports a substantial proportion of a commodity traded in the international market, or when the project under consideration has a significant impact on the world supply. As the import of the good increases to meet the input demand of the project, the CIF price may rise. Since the price increase will normally apply to a country's total imports, the relevant import price for the project input will lie above the new CIF price. In such cases, the import cost for the project, called the marginal import cost (MIC), equals the extra amount that must be paid for the total import divided by extra units purchased. This may be illustrated with reference to the example given in table 5.1. Suppose the initial import without the project is 100 units at CIF price of ₱3,000. With the project, the import goes up to 120 units and the CIF price increases to ₱3,250. Then MIC applicable to the project would be ₱4,500 as shown below.

Extra Cost = (120 x ₱3,250) – (100 x ₱3,000) = ₱90,000
Extra Units Imported = 20
Marginal Import Cost = ₱90,000 ÷ 20 = ₱4,500

To arrive at the economic price of additional imports, the border value of transport and distribution margins should be added in the manner shown in table 5.1.

The procedure for estimating the marginal export revenue (MER) resulting from the export of a project is similar to MIC's. In case the additional export contributed by the project leads to a lower price for the entire output, the MER for the project per unit of output will be below the existing FOB. Before taking up a project which affects the MER, a careful assessment of the world supply and demand situation should be made, and a project should be taken up only if it is justified by its MER.

Economic Price for Nontraded Goods

A commodity is nontraded either because by its very nature it cannot be exported or imported or its domestic price lies between the CIF import price and the FOB export price. Examples of the former are transport and other services, electricity, bulky goods with high transport costs, and goods which cater to the specific needs of a local community. A good may also be classified as nontraded when, because of special protection in the form of trade quotas or prohibitive tariff, no further imports are allowed into the country.

The valuation of a nontraded good in economic prices tends to be more complex than the valuation of traded goods and services because production or use of nontraded goods often affects domestic prices. The changes in prices, in turn, influence the use and production of these goods by other users or producers. If the supply of nontraded goods is elastic, its economic price is generally measured by the cost of supply, with all inputs valued at border prices. However, there may be instances where the supply of a good is fixed and its use as an input by the project affects the supply to other users. In such a case, the input's economic price should be derived from its marginal value to other users or its demand price. Valuation of these two types of inputs, as well as nontraded outputs, will be discussed briefly.

Marginal Supply Price

When the demand for a nontraded project input is met through incremental production, its economic price can be measured by the sum of all inputs (valued at border prices) required to produce the extra output. For instance, the economic price of electricity could be measured by the total cost—capital and current—to produce extra units of electricity required in a project. The required inputs will consist of both traded and nontraded components. The traded components, such as capital equipment and fuel oil, will be directly valued at border prices. The major nontraded components, such as building and site construction and other expenses, will be further decomposed into their traded and nontraded components. The traded components, as already explained, will be in border prices. The nontraded components, which will remain after the second round of decomposition, are likely to be small and should be border priced using the conversion factors discussed in chapter 8.

Marginal Value of Reduced Consumption

When a good is in fixed supply, its use as an input in a project will reduce its availability for other uses in the economy. Therefore, the price which other consumers are willing to pay, or the demand price, is the appropriate basis for determining the economic price of that good. Examples of such commodities are likely to be few. These would arise when domestic output cannot be increased either because of the lack of price incentives or controls by the government, or because essential inputs are not available due to import quotas or prohibitive tariffs, or when their domestic supply is fixed.

A project that uses only a small fraction of the total fixed supply of a commodity can reasonably use existing market prices as a measure of the demand price, which should be converted to its border price equivalent through the use of an appropriate conversion factor. However, if the demand for a good in fixed supply is large enough to change its market price, neither the old price nor the price after meeting a project's demand is a correct measure of the value of foregone consumption: the true value will be the value between these two prices. If the demand curve is linear and the change in price is small, it would be reasonable to take a simple average of the two prices as shown in the following equation:

$$P_m = \frac{P_1 + P_2}{2} \qquad (3)$$

Even for products that have nonlinear demand curves, the average of two prices can be a useful approximation—if price changes are not too large. Once P_m has been estimated, it should be converted into its border price equivalent by using the relevant conversion factor.

Valuation of Nontraded Outputs

Nontraded outputs are goods and services which are neither exported nor imported. The general rule for valuing these goods is to estimate the price which consumers are willing to pay for them. However, in case of commodities which are freely traded and whose production constitutes a small fraction of total supply, market prices can be used as a measure of economic value.

The willingness to pay assumes particular importance in situations where the above assumptions do not hold, either because the project output is large enough to affect the price of the product or where the government fixes the price below production cost and supply is rationed. In the former, neither the price without the project nor the reduced price after project implementation will serve as a true measure of consumer's willingness to pay. An average of the two prices, based on demand elasticity, will provide a realistic basis for valuing such goods. Valuation is more difficult for rationed goods and judgment has to be exercised to estimate the price which consumers would be willing to pay in the absence of rationing.

Pricing Issues in Specific Sectors

The above-mentioned principles for pricing inputs and outputs can be applied to most of the sectors and subsectors of an economy. The manner in which they are applied would, however, depend on the characteristics of individual projects. There are certain sectors with special characteristics that make the analysis of projects more complex. The following discussion illustrates the application of general principles and exemplifies the special characteristics and problems in certain sectors.

Minerals, Oil, and Natural Gas

The nonrenewable nature of minerals, oil, and natural gas differentiates these resources from those of other sectors. Since the availability of these resources in a country is known,[1] the pace of their use depends on the discretion of policy makers. When these resources are depleted, further use of the commodity in question can only occur through imports or through the use of domestic or imported substitutes. This raises special issues about the pricing of these products, which are examined below with reference to natural gas.

Valuation of Benefits

If a project involves the production of natural gas, its economic valuation depends on whether a country is a net importer or a net exporter of fuel. If a country is a net importer, the value of the output is the import price of gas or of the substitute displaced by the increased domestic supply of gas due to the project. The comparison of values should be made at the point of use on the basis of physical properties such as the calorific value of gas and its imported substitutes, with adjustments for differences in quality and convenience of use. If, however, the country exports gas, the value of output computed at the wellhead is equivalent to the FOB price minus the transport and handling costs to the port. In both cases, changes in the real price of oil and gas expected during the life of the project should be taken into account (see chapter 7).

Valuation of Cost

The complexity of analysis in estimating production cost arises because of the exhaustible nature of this resource. The economic cost of gas production should include the investment required for production and distribution and allowance for depletion of these reserves. The level of depletion allowance in economic analysis is determined by the switching cost, i.e., cost of replacing natural gas with alternative energy sources at the time these reserves run out. The switching cost, thus, depends on the size of known reserves of natural gas in relation to the timeframe within which existing resources would be used with and without the project under study. Use of natural gas in the current period brings closer the day of exhaustion and the need to switch to alternative sources. The opportunity cost of natural gas in the current year is thus the present value of the cost to the economy of the adjustment that must be made in the future. Present value is derived by discounting the cost of the substitute in the year in which reserves would be exhausted. A discount rate of 10–12 per cent is normally applied to arrive at the present value; a higher rate would correspondingly reduce the present value of the depletion premium. The calculation of the depletion premium is illustrated in the Annex to this chapter.

Pricing of Natural Gas as an Input

Natural gas is often used as an energy input, e.g., for power generation, for manufacturing fertilizer, or for fueling an industrial plant. The economic price of gas as an input in these activities should not be based on the cost of producing gas. Rather, it should be calculated according to the opportunity cost principles noted earlier. For instance, if the output of gas in a country can be used to replace the import of oil products used in the production of, say, electricity, then the economic price of natural gas should be the price of the oil products it replaces. If the country is an exporter of gas, the economic price of gas is its export price net of the transport cost to the point of export. However, in no circumstance should the economic price of gas as an input be below the economic cost of gas which includes production cost, distribution cost, and the depletion premium.

Fertilizer

In the case of a fertilizer project, the general principle discussed earlier can be readily applied. The value of a project output should be measured by the world price of fertilizer. If the country is a net importer of fertilizer, the economic price is the import price (CIF). If the country is a net exporter, the economic price is the export price (FOB), both adjusted for trade and transport margins. On the cost side, the economic price of the input depends on its source. If the production is based on imported naphtha, for example, the CIF price of naphtha is the economic cost of the input. If domestic gas is the input of the fertilizer plant, the opportunity cost of gas is the input cost.

In determining the economic price of fertilizers used as an input in a project, the relevant price is not the cost of production of fertilizer but the FOB price of fertilizer (if the country is a net exporter). If the country must import fertilizer to meet the needs of the project, the CIF price is used. Again, both FOB and CIF prices should be adjusted for trade and transport margins. The rationale behind this method of valuation is that if fertilizer were not used in the project, the country would have earned foreign exchange through exports in the first case and saved foreign exchange in the latter case.

Agro-Industries

One of the distinguishing features of most commercial crops is that they can be marketed only after processing. It is thus difficult to distinguish the economic value of a commercial crop from that of its processing. Furthermore, both the production and international prices of these crops fluctuate widely, making it difficult to predict future prices, and hence the economic rate of return. This is particularly true of tree crops, in which investments involve a gestation period of several years and yields continue for several decades. (For further discussion on price changes, see chapter 7.)

While the pricing of agro-industrial products poses some problems, the general approach mentioned earlier can be applied with minor modifications. To illustrate: let us take an oil palm project. The benefit of the project can be measured by the quantity and price of output—the price depending on whether the country is a net exporter or importer of edible oils. On the cost side, the cost of growing oil palm and the annual processing of its produce should be treated as a single project. If production and processing are treated as separate projects, the benefits of individual components will greatly depend upon the price paid by the factory to the growers for their produce—which may not necessarily be equal to the real economic cost of growing palm in the project area. The economic cost of the combined project should include the initial investment cost of growing palm—e.g., cost of land development and planting trees, recurring cost of inputs such as fertilizers, labor, irrigation, and the cost of transporting fruits to the factory, and investment and operating cost of the factory. However, where landholdings are small and individually cultivated, and the capacity of the processing plant does not match the output from the project area, it may be necessary to separate the economic evaluation of cultivation and factory operations. But the arbitrary nature of the decisions made in evaluating these separately should be explained.

Valuation of Land

Land valuation is generally omitted from textbooks on economic analysis of projects. This is perhaps because land is a renewable resource and, therefore, may be considered to have no inherent value. However, just like the other factors of production, land use has an economic cost and this cost should be measured by what it could produce in its next best alternative use. This would represent the true economic cost of the land used in the project which, like the other factors of production, should be expressed in border prices through the use of an appropriate conversion factor.

Land valuation in urban projects is more complex than in rural projects. In urban areas, land could be used for many purposes. Its value greatly depends on its location, availability of transport, and development of utilities like electricity and transport. If there is a free market, the market price or rent should provide a reasonably good measure of the value of land in its next best use. The problem, however, becomes complicated if the government fixes low prices for specific purposes like the development of industrial estates or low-income group housing. In such cases, demand outstrips supply by a wide margin and the government-controlled price does not provide a correct indicator of the true economic cost of land. In this situation, the demand price should be based on the cost of land with similar characteristics elsewhere within the same city or in another city of the same size.

The problem is more manageable in agricultural or mineral projects. In agricultural projects, the cost of land is measured by the stream of net benefits foregone as a result of taking out land from one use and putting it into another use. This is implicitly achieved in agricultural projects by defining project benefits as the difference between the "with-" and "without-" project situation. For instance, let us take the case of an irrigation project whose example is given in chapter 13. The benefits of the project area are measured by the difference between net output of the project with irrigation *minus* net output of the project without irrigation. In determining output without the project, one should normally take into account the likely change in output that may result from changes in institutional arrangements and cultivation practices. However, where no significant future changes are anticipated, the existing output without the project could be used as a proxy for the without-project situation.

In the case of a mineral project, the analysis of the opportunity cost of land is somewhat more complex. If the existing land is under commercial forest, then the loss of forestry benefits should be deducted to arrive at the net benefits of the mineral project. Also, as already noted, minerals are an exhaustive resource and, once a mine is exhausted, the country may have to depend on imported minerals for most of its domestic needs. Therefore, the present value of the imported mineral when the domestic resource runs out must be included in the cost of mineral output.

Endnote

1. This refers to natural endowments. Known reserves will increase as new discoveries are made.

ANNEX TO CHAPTER 5

Method of Calculating the Depletion Premium for Natural Gas

Estimating the depletion premium depends on:

- The number of years from the base year after which the gas reserves will run out. This requires estimates of total reserves, current output, and annual increase in consumption.
- The cost of alternative fuel which will have to be imported when gas reserves run out.
- The discount rate for estimating present value (PV).
- The energy equivalence between natural gas and alternative energy.

The method of calculating the depletion premium may be illustrated by taking the following example. Let us assume that:

(a) Country X is producing 90 million cubic feet (MCF) of gas whose consumption is growing at 10 per cent per year. At that rate, the reserves will run out in 24 years.

(b) Furnace oil which will have to be imported as alternative fuel in the 25th year from today will cost $60 per ton in constant base year dollars.

(c) One ton of furnace oil = 6 MCF.

(d) The discount rate for the country is 12 per cent.

Based on the above assumptions, the present value of the price of furnace oil imported 25 years from today is calculated below:

PV of furnace oil = $60 x 0.059 (at 12 % discount)
= $3.54

6 MCF = 1 ton furnace oil

1 MCF = $3.54/6 = $0.56

Therefore, the depletion premium, which is equal to the present value of alternative fuel required to replace gas after 25 years, is $0.56 per MCF.

CHAPTER 6

Economic Price of Labor

Labor is the single most important component among nontraded goods and services. Hence, an appropriate procedure for valuing labor is important for the economic analysis of projects. The wages paid by a government entity or a large private firm are not determined freely by the forces of demand and supply. A variety of factors such as the structure of the market, government regulations and control, labor unions, and collective bargaining, and the lack of labor mobility influence the determination of wages especially in the urban areas. Consequently, the prevailing wages do not reflect the opportunity cost of labor or their economic price. Similarly, in rural areas, because of the seasonal nature of jobs and the existence of unemployment, the market wage rates do not serve as a good measure of the opportunity cost of labor.

The purpose of this chapter is to suggest an approach for estimating the economic price of labor in developing countries, where labor markets are generally segmented or distorted for reasons mentioned above. The economic price of labor or the economic wage rate (EWR)[1] is derived in two steps, as in the case of commodities and services mentioned in the previous chapter. The first step is to estimate the EWR in domestic prices and the second is to convert the domestic value into border prices in which all costs and benefits are measured. This chapter deals with the first step. The second step involves the use of conversion factors; this is treated separately in the next chapter.

A correct measure of the EWR and its use in the early stage of the project cycle contribute not only to the efficient use of scarce resources but also to the appropriate choice of technique for a project. For example, if the EWR in a country is found to be very low, then labor-intensive methods of construction would be much more economic than capital-intensive technology that may be suggested by prevailing market prices.

General Principles for Calculating the EWR

The economic wage rate is frequently defined as the marginal output of foregone labor in one area of the economy because of its use in a new project in another area of the economy. This definition is based on the opportunity cost principle, which has also been applied to the economic pricing of commodities and services. The additional demand for labor generated by a project will affect not only the labor market in the sector to which the project belongs but also the labor market in other sectors. In measuring the opportunity cost, one is not concerned with the immediate source of hiring a worker for the project, but with the ultimate occupation from which the worker is drawn. For example, a project involving the construction of a cement plant may draw some workers from an existing fertilizer factory. The fertilizer plant may in turn replace the workers, setting in motion a chain of hiring that may end when agricultural laborers leave their farms and enter into the city for new employment. Consequently, it is the farms and not the fertilizer plant which ultimately provide the necessary workers and constitute the next best alternative use of the labor being employed in the cement project. It is, therefore, essential to trace the impact of additional labor demand through interlinked labor markets for the purpose of identifying the sector or sectors where the final adjustment will take place.

When labor used in a project is withdrawn from a productive sector or industry, the output of that sector or industry will normally decline. The magnitude of this decline is a function of the marginal productivity of the labor withdrawn from the sector or industry. Thus, its effect on national output greatly depends on the type of skills required for the project and the market for those skills. The EWR will be lower if a particular skill is in plentiful supply, i.e., if a large number of persons with that skill are unemployed. It is, therefore, often necessary to use a set of economic wage rates, one for each skill and location rather than a single rate for the whole country.

Due to lack of information in developing countries, it is generally difficult to accurately calculate the marginal product of labor (MPL) withdrawn for use in the project. However, a close approximation of foregone MPL is the prevailing wage rate of the skill in question adjusted for the degree of unemployment in that skill. In the case of skilled workers, market wages may be taken to represent the MPL, as these workers are generally in short supply. In cases in which severe unemployment and underemployment is expected to persist, the EWR would be lower than the money wage. However, in these cases, factors such as seasonal fluctuations in the demand for labor, varying degrees of labor mobility, and possibilities of self-employment suggest caution in adopting a very low opportunity cost of labor.

As in the case of commodities, it is also important to look at the supply side. At a very low wage, people may prefer leisure to work. This has been the conclusion of several project studies undertaken for large projects in Asian countries where unemployment was reported to be high. The supply price of labor depends on several factors such as family income, cost of migration, value placed on leisure and other nonwage activities (such as fishing or house repair), and nature and duration of project employment. Further, the "reservation wage"—below which people may prefer to remain unemployed or underemployed—is influenced by custom or convention. Although the reservation wage is likely to vary across classes of labor and geographical locations, the EWR is likely to be higher than what a narrow interpretation of the opportunity cost of labor would suggest.

Estimation of the EWR

Since labor is not homogeneous and consists of various types of skills, the opportunity cost of labor varies across skills and sectors. This makes the estimation of the EWR rather complex. Further, the EWR is, to a large extent, location-specific, especially in rural projects. Thus, even in the same sector, the EWR is likely to vary with the size and location of each project.

To keep the analysis manageable, it is suggested that an estimate be made of the total number of workers required by a project both during construction and operation and divided into three broad categories: skilled, semi-skilled, and unskilled. In certain cases where the threefold division is not possible because of data limitations or the nature of the project, the workers may be divided into skilled and unskilled groups. Then the EWR may be separately calculated for each category. Some suggestions for calculating the EWR for each category are given below.

Skilled Labor

Most developing countries have a shortage of skilled workers. Therefore, it is reasonable to use the market wage as the opportunity cost (or economic wage) for this category of workers. In countries where skilled workers are in short supply and where there is clear evidence that expatriate workers would be employed under the project, then the salaries required to attract them to the country should be used in estimating their cost. In calculating the opportunity cost, benefits other than salaries and wages (such as housing and provident fund contributions) should be taken into account.

Semi-Skilled Labor

In estimating the economic wage rate for semi-skilled labor, a distinction should be drawn between the "protected" wage sector and the "unprotected" wage sector, particularly with respect to projects in urban areas. In the former, wages can be held above the market-clearing level by minimum wage laws, collective bargaining agreements, and the hiring policies of the government and large companies. Wage levels in the "protected" wage sector should not be used for estimating the opportunity cost of semi-skilled labor. The wage levels in the "unprotected" sector should be used instead. The "unprotected" wage sector, which is beyond the control of wage regulations and labor unions, more accurately represents the opportunity cost of labor in urban areas.

The supply of semi-skilled labor is likely to vary considerably across countries. In many countries, the supply of these workers is barely adequate to meet growing demand. In such cases, the market wage rate in the "unprotected" sector should be taken to represent the opportunity cost of labor. This is so because even if workers are withdrawn from the "protected" sector for a project, they will leave vacancies that will be filled from the "unprotected" sector, including recent migrants whose alternative urban employment would most likely be in the "unprotected" sector.

In countries where there is an oversupply of semi-skilled workers and where there is clear evidence of unemployment, the opportunity cost should be adjusted downward, taking into account the extent of unemployment in the skills concerned. However, care must be taken in adopting very low EWRs for semi-skilled workers because these workers can find employment as unskilled workers. The EWR of semi-skilled workers should thus not normally be below the "unprotected" wage which unskilled workers can receive in a project area.

Unskilled Labor

Most of the discussions on EWRs in cost-benefit analysis concern the opportunity cost of unskilled labor. A large number of developing countries suffer from a high degree of unemployment and underemployment with most of the unemployed being unskilled. In determining the EWR for such workers, the starting point, as in the case of semi-skilled workers, should be the "unprotected" wage for unskilled workers in the project area. The use of such wage is particularly important in urban areas; in rural areas there is generally no "protected" wage; most workers belong to the unprotected sector. In determining the EWR, it should be kept in mind that unskilled unemployed workers in urban areas can engage in such unorganized activities as peddling, hawking and shining shoes, and the income from these should be taken into account in estimating the EWR. In rural areas the unemployed may engage in cooking or gardening, or provide help in family farming, or undertake seasonal work in nearby industries or in construction projects. These alternative activities and the economic value placed on leisure (or disutility of effort) should be taken into account when estimating the EWR.

In valuing the EWR in the local currency, direct information on annual income of unemployed and underemployed workers, if available, could be used. Where such information is not available, the annual EWR of unskilled workers could be estimated by multiplying the "unprotected" wage rate by the estimated number of days of gainful employment per year and by adding to it a rough estimate of other income based on surveys done in the project area.

Basis of Estimation

The major task in estimating the EWR is to secure reliable data concerning:

- Regulated and unregulated (or open-market) wage rates for skilled, semi-skilled, and unskilled workers in a project area. Unskilled workers form a fairly homogeneous group, and the wage rate applicable to them is likely to be more or less uniform in a project area. For skilled and semi-skilled workers, if the number of skills involved and the range of wage levels are large, an average wage rate should be estimated in that project area.

- The degree, nature, and duration of unemployment and underemployment in and around a project area. Data regarding farm employment during peak and slack periods as well as nonfarm income earned over the year should also be carefully examined. These data are normally available in national censuses, labor force surveys, and household income and expenditure surveys. Independent surveys made in the project or surrounding areas, wherever available, may be used to confirm the estimates obtained from official sources.

It may not be difficult to estimate the economic wage rates in countries where reliable and up-to-date data are available. Where such data are not available, it may be necessary to perform limited surveys as part of the feasibility study for a project. In any case, most of the work in estimating EWRs should be done at a fairly early stage of the project cycle so that based on the real economic cost of labor, various technical and financial options of implementing the project could be assessed before a final choice is made.

Estimating EWRs is particularly important in projects where the wage component in the total cost or benefit stream is significant and where technologies exist in formulating labor-intensive projects. It may not be necessary to estimate EWRs in projects where the wage component is small relative to the other components. This is the case in projects that involve few and mostly skilled workers and where technological choices are limited. However, for projects where the use of EWRs are necessary, they should be applied at early stages of project processing, even though the available data limit the estimate to a rough approximation and may call for further refinements later.

A convenient and useful approach in ascertaining the need for such refinements is to subject EWRs to sensitivity tests (see chapter 10). If the tests show that the economic internal rate of return (EIRR) of a project is relatively insensitive to changes in wage rates, further refinements may not be necessary. If, on the other hand, the EIRR is found to be relatively sensitive, further refinements may be warranted or justification may be required to show that the EWRs used provide an appropriate measure for the opportunity cost of labor.

An Example of Calculating EWR for a Rural Project

An example is given here for calculating the EWR for unskilled workers in a project which hires a large number of such workers for the construction of irrigation canals, rural roads, etc. The source of employment will be the rural labor force. Let us assume that the project area has a large pool of landless workers, more than enough to meet the needs of the project. To estimate the EWR of these landless workers, it is necessary to know their average daily income—based on the pattern of their productive activity during a year in which a worker is normally expected to work 250 days to be entitled to full wages.

Let us assume that, on an average, the landless workers were performing the following productive activities before being employed on the project.

Activity	Average Daily Income	Period
Planting and Harvesting	₱30	4 months
Fishing	₱25	3 months
Kitchen Gardening	₱20	3 months
Others	—	2 months

Based on the abovementioned activities, the opportunity cost of a landless worker hired for an irrigation project would work out to be ₱21.25 per day as shown below:

$$\begin{aligned} \text{EWR} &= 30\ (0.33) + 25\ (0.25) + 20\ (0.25) + 0\ (0.166) \\ &= 10 + 6.25 + 5 + 0 = ₱21.25 \end{aligned}$$

In estimating the above EWR, it is assumed that the worker attaches zero value to the two months which he spends in apparent idleness. It would, however, be incorrect to do that, especially if he has to move to a new job requiring full-time work. There are two reasons for this.

First, he may do odd jobs like repairing his house or helping other family members in activities which cannot be expressed in monetary values. Second, the worker may prefer leisure if the daily income falls below a certain minimum level. Therefore, he must be compensated adequately if he has to forego the two months of apparent idleness. The minimum payment required to persuade a worker to undertake productive activity is called the economic value of disutility of effort. If it is assumed that at least ₱15 per day is required to persuade the worker to forego idleness, the EWR of the landless workers adds up to ₱23.75:

$$\begin{aligned} EWR &= 30\,(0.33) + 25\,(0.25) + 20\,(0.25) + 15\,(0.166) \\ &= 23.75 \end{aligned}$$

The EWR is generally expressed in terms of a ratio of the economic wage rate to the market wage rate paid by a project. This facilitates the conversion of financial costs into economic costs. For this purpose, the average daily wage rate paid by the project to unskilled workers should be known. If the daily wage rate is ₱35, then the EWR ratio works out to 68 per cent on the basis of the following equation:

$$\text{EWR Ratio} = \text{EWR} \div \text{Labor's Market Wage Rate} \qquad (4)$$

or

$$\frac{₱23.75}{₱35.00} = 0.68$$

Therefore, to convert the financial cost of labor into economic cost expressed in domestic prices, the market wage bill for unskilled workers should be multiplied by 0.68. This means that the economic cost of labor is nearly two-thirds the financial cost of labor calculated for the project. It should be noted that the EWR that has been estimated in the above example is expressed in domestic prices, which must be converted into border prices on the lines suggested in chapter 8.

Endnote

1. In the literature on the subject, economic wage rate is also called the shadow wage rate (SWR). However, since in chapter II SWR is used as an abbreviation for social wage rate, the EWR will be used throughout this chapter for clarity.

CHAPTER 7

Treatment of Price Changes

General Approach

Prices of goods and services are the principal elements through which the benefits and costs of projects are measured and compared with to determine their economic or financial viability. All prices are, however, subject to fluctuations, with some changing faster than others. In the short run, prices are greatly influenced by the forces of demand and supply, but in the long run, cost of production, availability of raw materials, changes in tastes, and technology have more fundamental bearing on the prices of products and services. Price changes may also result from the implementation of government policies and adjustments in exchange rates which alter the relationship between traded and nontraded goods.

As mentioned in chapter 5, the economic costs and benefits of a project are normally expressed in terms of the prices prevailing in the year of appraisal. Considering the historical trend, the general price level can be expected to show an increase over the life of the project. Should this price increase be incorporated in the economic costs and benefits of the project? If the future trend in prices can be reasonably and accurately estimated, and the changes are consistently applied to both cost and benefit streams, the economic analysis can be equally done in current prices prevailing in each year of the project's economic life.[1]

The basic reason why economic prices are expressed in terms of prices prevailing in the year of appraisal is that it is virtually impossible to make any reasonable prediction of the likely changes in general prices, especially in the case of public sector projects where the economic life of the projects is very long. If the nominal prices of all commodities change at a uniform rate, the relative prices will remain unchanged and, therefore, no real price change has occurred. Since economic analysis ignores the transfer of resources from one section of the society to the other, a general price increase has no impact on the economic viability of the project.

Inflation and Financial Performance

A general price increase can, however, seriously impair the financial performance of a project in a number of ways. For instance, if a large cost overrun takes place because of a high inflation rate, and if the project entity fails to mobilize the required additional amount through equity or loans, project completion could be delayed and consequently complicate the problems of the project. A liquidity crunch at this stage could lead to insolvency. The liquidity problem could also arise due to the high cost of raw materials, spares, and finished goods in stock and other inventories. This could place the project at a serious disadvantage vis-a-vis the already established enterprises producing the same product. Consequently, the project which was initially expected to be a financial success could be a failure.

Inflation could impair the financial performance of a project through high interest rates which generally prevail during inflationary periods. Even if the project entity is able to borrow additional funds to finance the higher cost of the project, inflation may impose a high burden of interest payment and loan repayment, which the project may not be able to finance from its current income. This may lead to an erosion in the assets of the project entity, thereby impairing the financial viability of the project. This, in turn, could seriously affect the economic viability of the project, especially if the full output of the project cannot be realized due to financial problems.

Therefore, if after the financial appraisal of the project there is a considerable increase in the prices of inputs and outputs as well as increase in interest rates, the project should be reassessed for its financial viability. If necessary, the scope of the project should be further reviewed. This is essential even if the economic viability of the project appears unaffected as a result of inflation.

Relative Price Changes and Economic Viability

While a general price increase has no effect on the economic viability of a project, the situation can greatly alter if there is a change in the relative prices of inputs and outputs. This is because a change in relative prices changes the claims of the different products on the resources of a country. For instance, if the relative price of fossil fuels goes up by 50 per cent compared with the export price of wheat in a country, then the country has to part with 50 per cent more of its real resources through higher wheat exports to get the same amount of oil.

A study of the price trends of the various categories of goods over the past three decades shows that prices of commodities do not always move in the same direction: some move faster than others. As a result, the relative prices of commodities undergo changes, sometimes by a wide margin. A change in the real price should be measured in relation to the general price level prevailing during the same period. When changes in the price level are measured over several years—as in project analysis—the reference price must remain "constant" or "fixed" in order to obtain consistent results. Thus, the recommendation that costs and benefits should be measured in constant prices implies that changes in prices should be measured with reference to prices prevailing in the year of project appraisal. In this context, the term "base year prices" refers to average nominal prices prevailing in a particular year used for reference when measuring real price changes in the succeeding years.

In studying the impact of relative price changes on the economic analysis of projects, a distinction should be made between the capital cost and benefit sides of the project. Capital costs are generally incurred in a short span of two to four years, with the exception of some complex projects like multipurpose dams and integrated rural development projects.

In estimating total capital cost—which essentially involves expenditure on capital equipment, machinery, and construction—a reasonably accurate estimate can be made by using the prices prevailing at the time of appraisal and by making provisions for contingencies based on past experience in similar projects. In fact, while placing orders for the procurement of various investment goods, suppliers can be asked to quote firm prices expressed in a single currency (e.g., U.S. dollars) for deliveries to be made at specific times in the future. Therefore, the impact of relative price changes on the capital cost of the project is not likely to be significant. Other factors such as delays in project completion, inappropriate design, or unsuitable supply of equipment are likely to have a great impact on the economic cost of the project.

The impact of relative price changes could, however, be considerable on the benefit side of the project. This is for two important reasons. First, benefits accrue over a long period after completion of the project—in many cases the period is as long as 20 to 30 years or even greater. Therefore, if there is a change in relative prices of outputs and inputs involved in realizing the benefits of the project, their total impact calculated in terms of net present value could be very large, making an otherwise viable project nonviable and vice-versa.

The second reason is more fundamental and has to do with the long-term trend of prices of primary commodities relative to that of manufactured goods. A long-term comparison of price trends of manufactured goods and primary commodities shows that while the prices of manufactured goods have shown a steady increase, those of primary commodities have stagnated and even declined. Even where prices have increased, the increase has been much less than that in manufactured goods. As a result, prices of primary commodities relative to manufactured goods have shown a steady decline. This has often placed developing countries which remain producers and exporters of primary commodities at a serious disadvantage vis-a-vis the industrialized countries.

Estimation of Price Changes

Making projections of changes in real prices of inputs and outputs is perhaps the most difficult and challenging task which an analyst faces in the undertaking economic analysis of a project. It requires not only a projection of world prices of the commodities in question but also a projection of the manufactured goods which are used as a basis for determining changes relative to the base year. It also requires an estimation of relative price changes between nontraded goods including labor vis-a-vis manufactured and primary goods.

The task of making accurate projections is beyond the competence and resources of most project economists. The relative decline in the price of internationally traded primary goods in the past was the result of several factors including large productivity improvements, discovery of new and cheaper sources of supply, extensive use of material saving technologies, changes in tastes, and, most importantly, massive doses of subsidies provided by many developed countries for increasing their agricultural production. It is nearly impossible to predict the extent to which these factors will continue to operate in the future and thereby keep the prices of primary commodities depressed.

It may be mentioned that the World Bank[2] undertakes biennially a major exercise (in real and nominal prices) of forecasting over 40 key agricultural commodities, minerals, and fuels. These constitute, by far, the most comprehensive and best prepared estimates undertaken worldwide. These forecasts extend over a period of 10 to 15 years and serve as a useful basis for price projections for relevant commodities. Many international institutions make use of these projections in estimating the benefits of projects in the agricultural and minerals sectors.

Despite the use of large resources of manpower and sophisticated analysis, the World Bank projections have not proved very reliable. In fact, an analysis of the projections made in the past 12 years shows that there was a wide divergence between the projections and the actual prices. The projections themselves had undergone large revisions in relatively short periods of time. This is not an adverse commentary on the analysis carried out by the World Bank but indicates the difficult nature of the task involved.

Given the limitations inherent in projecting future prices, and the fact that the World Bank projections are limited to some agricultural, mineral, and metal products and do not cover a large number of internationally traded and nontraded goods and services, it is not possible to use them across sectors. It is therefore suggested that, as a base case, real prices prevailing in the year of appraisal should be used in all projects. The underlying assumption in such analysis is that the relative prices prevailing in the year of appraisal will continue in all future years, and that if there are any year to year variations, they will get canceled over the life of the project. This is not to suggest that the base year prices should not be adjusted under any situation. If, based on the past trend, the base year prices are considered unduly high or low, an average price for the recent three to five years could be used.

The abovementioned method of using base year prices should be used for all projects for the sake of consistency. But in the case of commodities for which World Bank projections are available, the project viability should also be tested using those prices. Such a test could be applied separately or as part of the sensitivity analysis (see chapter 10). If the difference in the two rates of return is very large, the implications should be clearly spelled out so that policy makers can take an informed decision on the matter. If the project becomes nonviable using the World Bank projections, the analyst is expected to give his assessment of the price projections in his presentation of the alternatives before the policymakers.

Endnotes

1. However, in this case the discount rate used in estimating the project's net present value or the economic rate of return should also include the rate of inflation prevailing in each year.

2. See World Bank, *Price Prospects for Major Primary Commodities* (November 1988). These forecasts are prepared by the World Bank every two years and are updated at a 6-month interval to take into account the emerging trends. The latest estimates which adopt 1985 as the base year extend up to the year 2005.

CHAPTER 8

Use of Conversion Factors

In chapters 5 and 6, the valuation of goods and services and labor was discussed. While traded goods and services have been valued in terms of their prices at the border, the value of nontraded goods and services and labor has been expressed in domestic prices.

This chapter suggests a set of conversion factors at different degrees of disaggregation which will help convert the domestic values of nontraded project components into their border price equivalents. This, together with the traded components, will result in expressing the total cost and benefits of the project in border or international prices. This system of measuring all costs and benefits in border or international prices assumes that international trading opportunities open to a country provide the basis for calculating the economic worth both of domestic output and of factors of production. The unit of account (or numeraire) is the "present value of public income in terms of foreign exchange". This means that the real values of imports and exports (CIF and FOB, respectively) become the reference prices on which production decisions are made.

Types of Conversion Factors

A conversion factor is the ratio of a border price of a commodity to its market price. By applying this conversion factor to the domestic market price of a commodity, its border price (or equivalent) can be determined. In principle, a separate conversion factor should be estimated for each nontraded commodity. However, in practice, this is seldom possible and, frequently, a common conversion factor is calculated for a group of commodities. Generally, three types of conversion factors are distinguished: (1) commodity conversion factors (CCFs) for specific commodities or services; (2) group conversion factor (GCF) for a group of related commodities and services; and (3) standard conversion factor (SCF) for all commodities and services produced (or consumed) in the country. These conversion factors are briefly discussed below.

Commodity Conversion Factors (CCFs)

CCFs are used for converting the values of specific commodities which are used as inputs or produced as outputs in a project into their border price equivalents. Since by their very nature the commodities do not enter into international trade, these CCFs are estimated through a decomposition process, illustrated later in this chapter. The use of such CCFs is generally restricted to a few major nontraded inputs and outputs encountered in a project.

The major advantage of the CCF approach is that it corrects distortions specific to each project. For example, the principal nontraded element in one project may be electricity, for which the conversion factor may be 1.5; for another project, the principal nontraded element may be road transport, for which the conversion factor may be 0.6. If the nontraded components of both projects are converted not with these specific conversion factors but with a single SCF of, say, 1.2, the real resource cost of electricity at border prices will be understated in the first project, while that of road transport at border prices will be overstated in the second project.

Group Conversion Factors (GCFs)

A GCF is required when project costs and benefits cannot be specified readily in terms of individual commodities and services, but only in terms of aggregates of such commodities and services. In the cost estimates, these aggregates may, for example, be the cost of civil construction, domestic machinery, and transport and distribution margins. The analysis of these costs in terms of individual commodity components may be time-consuming and difficult. To avoid this, a set of GCFs is estimated to be used for adjusting costs and benefit items common to several projects.

Standard Conversion Factor (SCF)

The SCF may be thought of as a type of GCF applicable to all commodities consumed or produced in the economy. It is the average of all the discrepancies between domestic and border prices. If commodity-specific and GCFs are not available, then the cost of some major nontraded items may be decomposed into their traded and nontraded components. While the traded components could be directly border-priced, the SCF could be applied to the nontraded components of those items, as well as all remaining nontradable goods, to convert their aggregate cost into equivalent border prices.

In cases where even data required to carry out the above-mentioned adjustments are not feasible, the SCF may be applied to the entire cost of nontraded items without any adjustments. In such cases, the conversion factor approach is equivalent to the shadow exchange rate (SER) approach, the only difference being that the former expresses costs and benefits of a project in terms of border prices, while the latter expresses them in terms of domestic prices. The SCF indicates border prices of goods worth one unit when valued at domestic prices, while the SER indicates the domestic price of goods worth one unit when valued at border prices. For any economy, the SCF factor of 0.8 is equivalent to an SER of 1.25 (i.e., 1.0/0.8).

The SCF and SER do not measure the extent of foreign exchange imbalance, only the average discrepancy between domestic and border prices. For example, a country's balance of payments may be in equilibrium at the prevailing exchange rate, yet the SER as defined above will still differ from the prevailing rate if there are distortions due to factors such as duties and quotas. This is true for any country irrespective of whether its balance-of-payments position is favorable. It may be mentioned that the SCF and SER only attempt to compare values expressed in domestic prices with values expressed in border prices. They should not be regarded as indicators of the equilibrium exchange rate or the premium that may be attached to the foreign exchange in a situation where foreign exchange is scarce.

Conversion Factors for EWR Calculation

In chapter 6, the EWR was defined and it was indicated how its estimate could be derived in a local currency. The estimated value must be converted into border prices, to be added with other costs and benefits items. To do so, it is essential to identify the source from which the labor is withdrawn for the project. For instance, if skilled labor is withdrawn from, say, a fertilizer plant, then the value of output ascribed to the industrial worker could be expressed in border price through the use of a conversion factor for fertilizer. Similarly, unskilled labor may be ultimately drawn from rural areas where foregone output may be rice and corn. In this case, the conversion factor derived from the weighted average of rice and corn could be applied to the domestic EWR to arrive at the border price of the opportunity cost of unskilled labor. In situations where the sources of labor cannot be identified, the SCF could be used to determine the border price of labor.

Method of Calculation

Having identified the types of conversion factors, the method of calculating these conversion factors from the available data will be briefly discussed. The CCFs will be discussed in some detail because the decomposition method of CCFs is equally relevant to GCFs. Conversion factors for labor will not be discussed further because such factors take the form of either CCF or GCF.

Conversion Factors for Nontraded Inputs

Before applying a specific commodity conversion factor, the domestic economic price of the nontraded commodity has to be estimated first. As shown in chapter 4, most of the inputs are likely to be supplied through increases in production. In such cases, the economic price should be based on the increased cost of supplying the input. In this context, the example of electricity—which is a common and important nontraded input in most projects—is taken up and it is shown how its incremental cost of supply should be determined.

Suppose that a project requires electricity which is supplied by building a new thermal generating plant. The economic price of electricity is the value of all inputs expressed in border prices used to produce electricity. Table 8.1 gives the relevant details, including the capital and annual operating cost of producing one megawatt of electricity from a thermal generating

TABLE 8.1
Border Price of Electricity Valued at Marginal Cost of Supply
(Thousand Pesos)

Item	*Cost in Domestic Prices*	*Conversion Factors*	*Cost in Border Prices*
Capital Cost	15,000	0.9	13,485
Thermal Generating Unit (CIF)	9,000	1.0	9,000
Building and Site Construction	6,000	0.75	4,485
Imported Materials (CIF)	2,500	1.0	2,500
Labor	1,500	0.7	1,050
Taxes and Duties	1,000	0	0
Other Expenses	1,000	0.93	935
Imported Materials (CIF)	750	1.0	750
Labor	150	0.7	105
Others	100	0.8	80
Annual Operating Costs	5,000	0.85	4,260
Fuel Oil (CIF)	4,000	1.0	4,000
Maintenance	1,000		260
Parts (CIF)	150	1.0	150
Labor	100	0.7	70
Others	50	0.8	40
Taxes and Duties	700	0	

plant. The steps involved are (1) to decompose both capital and operating costs, (2) to express them in border prices, and (3) to translate them into cost per kilowatt-hour.

As shown in table 8.1, the total capital cost of the power plant is ₱15 million. This is divided into thermal generating unit, which is entirely imported, and building and site construction. The CIF price of ₱9 million for the thermal unit is also its border price. The nontraded component of ₱6 million for building and site construction is decomposed into imported materials (traded), labor (nontraded), taxes and duties, and other expenses (a mix of traded and nontraded items). The last item requires another round of disaggregation as shown in table 8.1. The annual operating cost is similarly divided into fuel oil and spare parts (imported), taxes and duties (transfer), labor, and other items.

Table 8.1 shows that through the process of decomposition, the total cost of electricity can be divided into four parts; the largest part, constituting the traded component, can be directly border priced. Taxes and duties involve no economic cost and should be disregarded. The conversion factor for labor is assumed at 0.7. The residual items which constitute a small fraction of the capital and operating cost of the power plant could be border priced by using the standard conversion factor, which is assumed to be 0.8. If the residual element is fairly large, it could be further decomposed into its traded and nontraded costs. However, normally, two or three rounds of decomposition are adequate for carrying out the necessary corrections.

Based on the calculations made in table 8.1, the capital cost of the power plant in border prices is ₱13.5 million, compared with ₱15 million in domestic prices. The ratio of domestic to border prices works out to 0.9. Similarly, annual operating costs in border prices total ₱4.26 million, compared with the ₱5 million in domestic prices, which yields a conversion factor of 0.85.

Having estimated the total cost of producing one megawatt of power in border prices, it is necessary to express this cost per kilowatt-hour (kWh), a standard unit of measurement. To do this, it is first necessary to put the capital and operating cost on a common basis. Assuming that the operating life of the thermal power plant is 25 years, it is necessary to find its annualized capital cost to which should be added the annual operating cost to estimate the cost of generating electricity in one year. Since the present value of future investments is always lower, it would be incorrect to derive the annual capital cost by dividing the total cost by 25 years. The correct formula for estimating the total annual cost of generating electricity is given in the following equation:

$$\text{Annual Cost} = Pv \frac{q}{1-(1+q)^{-n}} + AOC$$

where

Pv = Present value of capital cost in economic prices,

q = Discount rate for public sector in economic prices,

n = Useful life of capital assets, and

AOC = Annual operating cost.

Based on the estimated life of 25 years and a discount rate of 12 per cent, the annual capital cost of producing one megawatt of additional electricity works out to ₱1.73 million.

$$\text{Annual Capital Cost} = ₱13.5 \text{ million} \times \frac{0.12}{1-(1.12)^{-25}}$$

$$= ₱1.73 \text{ million}$$

Adding the annual operating cost of ₱4.26 million, the total annual cost of generating one megawatt of electricity totals ₱5.99 million.

To obtain the cost per kWh, the capacity utilization of the power plant has to be estimated. Assuming that the capacity utilization is 75 per cent, then the total electricity generated during the year would be 6.57 million kWh (6,570 hours x 1,000 kW). On this basis, the cost per kWh would be ₱0.91. The total electricity requirement of the project should be valued at this price and not at the price at which the current consumers are getting electricity.

The example above provides a method of estimating the marginal cost in border prices of a nontraded input which is supplied through increased supply. As already noted in chapter 5, there may be some instances when a project uses as an input a good in fixed supply. In such a situation, the market price would serve as a measure of its domestic economic value. Since such nontraded inputs are likely to be few, in the absence of other data, the SCF could be used to estimate their border prices.

Conversion Factors for Nontraded Outputs

The previous example estimated the border price for electricity on the basis that it was a project input met through higher output. However, where the border price has to be estimated for the project whose output is electricity, a plausible assumption is that additional output of electricity would lead to an increase in the consumption of electricity in the economy.[1] An increase in the consumption of electricity leads to a diversion of funds spent on substitutes. For example, if 60 per cent of the additional expenditure on electricity results from a reduction in diesel consumption and 40 per cent from a reduction in coal consumption, and if the border price of diesel is twice its domestic price and the border price of coal is 1.5 times its domestic price, then the conversion factor for electricity as an output is 1.8 = (2 x 0.6 + 1.5 x 0.4).

Calculating conversion factors for projects whose output is nontraded, therefore, entails finding the weighted average of the conversion factors for closely related complementary and substitute commodities and services. Since it is difficult to trace all likely changes in the consumption of other commodities, in practice the conversion factors should be based on changes likely to occur in a few directly competing (or complementary) commodities or services.

In the case of infrastructure projects, the benefits are generally expressed in terms of cost saving. For instance, in a road project, the benefits may be savings in fuel costs, lower use of spare parts and tires, etc. In a port project, it may be savings in terms of freight charges. In such cases, the cost savings must be disaggregated into components and an appropriate conversion factor should be applied to each component to arrive at an overall conversion factor. Since there is a general stability in relative proportion of various components, the overall conversion factor, so derived, could be used for a number of similar projects over the years.

Group Conversion Factors (GCFs)

GCFs are defined as the weighted averages of the commodity conversion factors. It is adequate to use GCFs, except for some key commodities in a project. There are two major advantages in doing this.

First, the GCF may be based on weights derived from a more general body of readily available data and will be of use when project-specific weights are not available. Second, the GCF can be based on group-level data of the ratio of border prices to market prices, which may be easier to estimate than similar ratios of individual commodities. Therefore, it is likely that the calculation of the GCFs will be easier than the estimation of individual CCFs.

GCFs may be calculated only for commonly encountered aggregates. On the project cost side, these may include civil construction, domestic machinery, transport, and distribution margins. The EWR calculation in border prices may require the estimation of a GCF for industrial output, agricultural output, or for low-income consumption.

Each GCF should be calculated as a weighted average of the ratio of border prices to the market prices of its major components. This may be illustrated by taking the example of the GCF for civil construction. Let us assume that civil construction has only four components of steel, cement, wood, and labor. Their relative weights in total construction and conversion factors are given in table 8.2.

TABLE 8.2
Estimation of GCF for Civil Construction

Items	*Weights*	*Conversion Factor*	*GCF*
Steel	30	0.70	0.21
Cement	40	0.90	0.36
Timber	20	0.60	0.12
Labor	10	0.80	0.08
Total			0.77

GCFs can thus be estimated by averaging the conversion factors of the traded goods and nontraded goods using appropriate weights. For nontraded goods included in the weighing system for which conversion factors are not available, the SCF can be used. The weights for GCF for civil construction, domestic machinery, transport, and distribution margins can be derived from enterprise accounts or project reports. Weights for the GCF for agricultural output can be obtained from agriculture production statistics, and those for consumption may be based on available consumer expenditure surveys.

Standard Conversion Factor (SCF)

The SCF is a group conversion factor for which the group covers all commodities produced or consumed in the economy. Hence, it should be ideally calculated as the weighted average of the conversion factors for specific commodities or group of commodities. However, since this is not generally feasible, certain shortcuts can be used for calculating the SCF.

The commodities produced or consumed in the economy include both traded and nontraded commodities. The conversion factors for nontraded commodities could, in principle, be derived, similarly to the example of electricity in terms of traded commodities. As in the case of GCFs, a suitable weighing system should be established. Ideally, a weighing system derived from input-output tables may be used. However, since such data is not available for most countries, trade weights, which are more easily available, are normally used and the *SCF* is derived from the following formula:

$$SCF = \frac{M + X}{(M + Tm) + (X - Tx)}$$

where *M* and *X* are values of imports and exports in border prices, respectively, where *Tm* is revenue from import duties net of subsidies, and *Tx* is revenue from export duties net of subsidies.

The method suggested above may not be suitable if the spread between domestic and border prices is not fully accounted by border taxes and subsidies and trade and transport margins. This is the case where significant nontariff distortions exist. In such cases, effort should be made to adjust the above *SCF* for distortions observed for principal commodities.

The formula for calculating the *SCF* in the example given above is a rather crude approximation because (1) it assumes that the share of various commodities in the total value of trade approximates the shares in production (or consumption) and (2) it does not take into account the possibility that for some commodities the spread between domestic prices and border prices is greater than the net border tax.

A better approximation can be obtained by separating agricultural commodities from industrial commodities. A GCF for agricultural commodities can be calculated directly from data on domestic and border prices and the weights derived from agricultural production statistics. For industrial commodities, the GCF can be approximated by the border value formula. The *SCF* is then defined as the weighted average of these two GCFs:

$$SCF = W_a C_a + W_m C_m$$

where C_a and C_m are the GCFs for agricultural and industrial output and W_a and W_m are the weights attached to them, the weights being the relative share of agriculture and industry in total production.

Endnote

1. Such an assumption is tested through demand surveys in the project area before a project is undertaken.

CHAPTER 9

Comparison of Costs and Benefits: Investment Criteria

Having identified, quantified, and valued the economic costs and benefits of a project, the next step is to compare these two flows to determine whether the project would result in the efficient use of resources from an economic point of view. A project must satisfy at least two conditions to be acceptable for investment. First, it must yield benefits in excess of costs over its economic life. Second, the net benefit must be larger than, or at least as large as, that resulting from the next best alternative project. Otherwise, the implementation could lead to reduced national welfare or a lower contribution to national output than would be possible through the implementation of alternative projects.

Comparison of Project Alternatives

The comparison of mutually exclusive project alternatives is, therefore, an essential condition for the determination of the economic acceptability of the project. As explained in chapter 3, there are usually many projects or project options which by their very nature are mutually exclusive: if one is selected, the others cannot be undertaken. This applies to different sizes, design, technologies, and time-phasing of what is essentially the same project, and the choice of different locations for a project which meet the needs of a given market. Some projects are mutually exclusive in the strict sense that they represent alternative ways of producing exactly the same output (e.g., hydro versus thermal power generation). Since the benefit is the same, the comparison involved is only in terms of cost. Therefore, the option that should be selected is the one which has the lowest present value of cost when discounted by the opportunity cost of capital—which has a great bearing on the determination of the least-cost option (see chapter 3).

As noted earlier, the fact that a project constitutes the least-cost alternative among the feasible options does not, by itself, tell anything about the economic merits of the project as compared with the "without-project" alternative. This is because even the least-cost project may have costs which exceed benefits. Therefore, in such projects, the analysis should not stop at the determination of the least-cost solution but, wherever possible, consider whether benefits are adequate. In particular, differences in costs as between least-cost design and the next-best alternative are not, and should not be, regarded as a proper measure of the benefits of such projects.

Since the streams of investment costs and benefits occur at different time periods and investment costs are incurred over a short period compared with the benefits, they cannot be directly compared for estimating the net benefits of the project. A system of discounting is used to determine the present value of the costs and benefits of the projects. The discount rate is based on the opportunity cost of capital to the economy. After determining the net present value of the benefit and cost streams of the project, alternative sets of investment criteria are used to determine project viability. This chapter discusses the system of discounting, estimation of the opportunity cost of capital which serves as a basis for the appropriate discount rate, various investment criteria, and their appropriateness in determining the economic viability of a project.

System of Discounting

In all societies, greater value is placed on investment and consumption today compared with that in the future. This is equally true of money. The existence of the time value of money causes lenders to demand interest payment on their loanable funds and the borrower to pay such interest. This is analogous to the practice followed in traditional societies where a better-off farmer loaned a measure of grain on promise of the return of, say, one-and-a-half measure after the new crop was harvested.

What is true for an individual is equally true for the economy as a whole. Resources available today can be invested to produce goods and services which will be available to the country for many years to come. If the same resources were made available two years later, then the economy will lose the benefit of investments which would have been realized in those two years. The cost of waiting in this case would be equal to the economic value of goods and services produced by the economy had the resources been made available two years earlier.

Because of the time value of money, it would be inappropriate to equate the value of, say, ₱100 of a given benefit (or cost) realized in year one with that realized five years later. The process of weighing cash flows of cost and benefits according to the year in which they occur is called discounting; the percentage difference between the value of a peso now and its value a year from now is called the discount rate. Through the use of discount rate, all future costs and benefits are expressed in terms of their present value, or present worth of pesos earned or spent in future years. The process of discounting is thus simply compounded interest rate worked backward.

Let us take a numerical example to illustrate the use and meaning of the discount rate. Let us assume that a person has ₱1,000. He can earn an interest of 10 per cent by depositing it in a bank or buying a government bond. This means that capital of ₱1,000 will increase to ₱1,100 a year from now. In other words, the present value of ₱1,100 a year from now is ₱1,000. That is:

$$₱1{,}000 = \frac{₱1{,}100}{(1 + 0.10)} \qquad (1)$$

Let us now extend the period to three, assuming that the return is 10 per cent for each year. In this case, a person's 1,000 will be worth ₱1,331 three years from now. This can be expressed as follows:

$$₱1{,}000 = \frac{₱1{,}331}{(1 + 0.10)^3} \qquad (2)$$

Let us consider two variations of equation (2). Let us assume that the person is offered ₱1,331 to be paid after four years instead of three years. In this case, the present value of his investment at 10 per cent would be ₱909, and it would be unwise for him to make such an investment. As a second alternative, let us assume that the person is offered a repayment of ₱1,331 after three years on his investment of ₱1,000, when the bank interest rate is 6 per cent. In this case, the present value of ₱1,331 received at the end of three years would be:

$$\frac{₱1{,}331}{(1 + 0.06)^3} = ₱1{,}118 \qquad (3)$$

which is greater than the present value at 10 per cent.

The following three points emerge from the above discussion of the discount rate:

(1) The net present value of any sum is lower than its future nominal value.
(2) The longer it takes to realize a certain nominal value, the lower its present value.
(3) The lower the discount rate, the higher the present value, and vice-versa.

It is obvious from the discussion above that if one has to determine the net present value of the flow of costs and benefits spread over several years, as is the case in almost all public sector projects, then each of these flows will have to be discounted depending upon the year in which they occur. Also, since project costs are incurred in early years while benefits are realized in later years, the impact of a given discount rate is greater on benefits than on costs. Therefore, following point number (3), a project would be more viable if the discount rate is lower, while a project would be less viable if the discount rate is higher.

Opportunity Cost of Capital

The choice of an appropriate discount rate has a major bearing on the determination of the economic viability of projects. For public sector projects, the discount rate should represent the opportunity cost of capital to the economy, just as to the private individual it should represent the cost of capital raised from the market. While it is relatively easy to define what the discount rate in the latter case should be, it is very difficult to correctly estimate the real cost of capital to the economy. There is a great deal of academic debate on this subject, and various suggestions have been made as to how the cost of capital should be estimated.

Since the public sector in most countries is short of investible resources, heavy reliance is placed on borrowing from the domestic private sector for financing its investment program. If an economy is fully monetized and interest rates paid and charged by the banking system are freely determined by the forces of demand and supply, the market rate of interest could be used as a proxy for the OCC. In practice, however, this is far from being the case. First, the banking system is not adequately developed and caters mainly to the needs of the organized sector of the economy. Further, the banking system in most countries is either owned or controlled by the government, and the interest rate fixed by government authorities may not represent its true economic value. In fact, interest rates are often kept low to contain the debt-service burden of the public sector, and this depresses the saving rate in the country.

The discussion above looks only at the supply side of capital. To determine the real cost of capital, one should look at both its supply and demand sides. On the demand side of capital, the return is measured by what an investor can make on the borrowed funds. In an ideal situation where there are no taxes, inflation, and restrictions on interest rates, the equilibrium rate for demand and supply for funds will be the same. However, because of the existence of the factors mentioned above, there is generally a wide difference between the real return on investment and the interest rate paid by a bank.

Despite the limitation of data, estimates of the OCC have been made by taking a weighted average of the real return on investment and interest rate on fixed deposits with commercial banks. Generally, the average figure ranges from 10 to 15 per cent. Most international institutions adopt 10–12 per cent as the cut-off rates. This means that projects with a real rate of return of at least 10–12 per cent are considered economically viable. Projects with returns below 10 per cent are financed in exceptional cases and only where there are large benefits which cannot be valued in monetary terms.

Most developing countries finance a substantial part of their development expenditure through external assistance. It could be argued that a shortage of investment resources in these countries could be met by increasing the inflow of external resources. Since the volume of official development assistance, bilateral as well as multilateral, is subject to a system of informal country quotas, any additional marginal inflow can come only through commercial sources. Many developing countries are not creditworthy and may not be able to raise commercial loans. Others may have to pay a substantial premium over the prevailing international rate which varies from year to year. Also, there is risk of a revaluation of foreign currencies whose interest rates are low and of a devaluation of domestic currencies. Because of these considerations, the abovementioned OCC of 10 per cent appears to be the minimum rate which most developing countries should use as the discount rate in judging the economic viability of investment projects.

An alternative approach would be to treat the discount rate as a rationing device. All projects could be ranked on the basis of their rate of return. Projects with the highest returns which can be accommodated within the available resources could be implemented, while others could be deferred or dropped. This approach is feasible only when (1) the investment requirements of projects under consideration exceed the available resources with the government and (2) all the projects have been appraised and their net present values or rates of return have been estimated. However, serious difficulties may arise when an analyst has to assess the viability of projects whose existing prices have been artificially fixed through government regulation and control. It is, therefore, suggested that for purposes of economic analysis, a discount rate of 10–12 per cent should be used, as is the existing practice of most international lending institutions.

Alternative Investment Criteria

Several criteria have been used by economic analysts to judge the performance of projects and their economic viability. Four commonly used methods are: (1) net present value, (2) economic internal rate of return, (3) benefit-cost ratio, and (4) pay-out period. The first two approaches are most often used, singly or combined, to determine project viability. They will, therefore, be examined in greater detail. The last two methods are not commonly used for reasons that will be explained later.

Net Present Value

The net present value (*NPV*) is the most common criteria used in determining the economic soundness of projects. The *NPV* is measured by the difference between the present value of the benefit and cost streams of a project during its economic life, discounted by the *OCC*. If the *NPV* is positive, the project should be accepted; if it is negative, the project should be rejected. This is because the *NPV* measures the amount by which the economy will be made better or worse by the implementation of a project.

In the previous discussion of the discount rate, it was explained that because of the time value of money, the present value of the stream of future costs and benefits declines over time. The present value of all flows of net benefits $(B_0 - C_0)$, $(B_1 - C_1)$ .... $(B_n - C_n)$ can be expressed algebraically as:

$$NPV = \frac{(B_0 - C_0)}{(1+r)^0} + \frac{B_1 - C_1}{(1+r)^1} + \frac{B_2 - C_2}{(1+r)^2} .... + \frac{B_n - C_n}{(1+r)^n} \quad (4)$$

$$= \sum_{t=0}^{n-1} \frac{B_t - C_t}{(1+r)^t} \quad (5)$$

where r represents the *OCC*.

Note that the calculation of the *NPV* requires (1) an estimation of benefit minus cost for each year, (2) an application of the appropriate discount factor, and (3) the addition of the resultant values. An illustrative example is given in table 9.1 to clarify the point.

Let us assume that the economic cost of a food processing project is ₱12.000 million spread over two and a half years, while its benefits are spread over eight years beginning on the first half of the third year (as shown in table 9.1). The *NPV* can be calculated either by (1) estimating separately the present value of stream of costs (*C*) and benefits (*B*) and then arriving at the net figure by deducting *C* from *B* or more simply (2) deducting costs from the benefits in each year and then estimating *NPV* of *B–C*. In table 9.1, the second approach is used. Based on the calculations shown in this table, the *NPV* of the project is estimated at ₱3.845 million.

The *NPV* is particularly important in choosing which public sector projects to implement. Governments in practically all developing countries face resource constraints in undertaking in-

TABLE 9.1
Example of Net Present Value
(Million Pesos)

Year	*Benefits (B)*	*Costs (C)*	*B–C*	*Discount Factor at 12 Per Cent*	*Present Value*
0	–	9.000	–9.000	1.000	–9.000
1	–	2.000	–2.000	0.893	–1.786
2	2.000	1.000	+1.000	0.797	.797
3	4.000	–	4.000	0.712	2.848
4	4.000	–	4.000	0.636	2.544
5	4.000	–	4.000	0.567	2.268
6	4.000	–	4.000	0.507	2.028
7	4.000	–	4.000	0.452	1.808
8	4.000	–	4.000	0.404	1.616
9	2.000	–	2.000	0.361	.722
Net Present Value (NPV) Total 0 to 9					3.845

vestment projects. The objective, therefore, is to maximize the development impact of available resources. This can be achieved by implementing projects which give the highest *NPV* with the available investible funds. To clarify the point, let us assume that the agricultural ministry in a small country has a budget of $10-million equivalent and has the following four projects to choose from:

Project A costs $3 million, *NPV* $0.2 million
Project B costs $7 million, *NPV* $2.8 million
Project C costs $4 million, *NPV* $1.0 million
Project D costs $6 million, *NPV* $1.5 million.

With the budget limit of $10 million, projects A and B should be chosen because they give higher combined *NPV* than projects C and D—$3.0 million as against $2.5 million. Despite the fact that project A has the lowest *NPV* per unit cost, it should be selected because together with project B, it maximizes the *NPV* for the country. If, however, the budget limit is increased to $11 million, projects B and C should be preferred because with just a $1 million increase in investment, *NPV* increases by as much as $0.8 million.

The *NPV* method is effective only when three conditions are present.

First, there must be a large number of fully appraised projects available with the government, whose total cost exceeds by a significant margin the available budget; otherwise, less desirable projects might be taken up in place of better ones which are still to be appraised.

Second, the investment budget for the government or the concerned ministry must be fairly firm; otherwise, the ranking of projects can change.

Third, and most important, the *OCC* should be known and fairly firm. This is because the choice and ranking of projects in terms of their *NPV* can change completely with the change in the discount rate. For example, with a high discount rate, projects which yield high benefits in the earlier years would be preferred while with a low discount rate, projects with high benefits in later years would appear preferable.

Another constraint in using the *NPV* is this: the *NPV* represents a surplus generated by a project over its economic life in excess of what it would be if investments were used elsewhere in the economy.

Since the *NPV* has a time dimension, it would be incorrect to compare two projects which may involve the same cost but have very different economic lives. For instance, it would not be meaningful to compare the *NPV* generated by a project with an economic life of 30 years and costing, say, $1 million with the *NPV* of a project which costs as much but with an economic life of five years.

When the project life is very different, a comparison could be made by repeating the project with the shorter life as many times as required to make its life comparable to that of the other project. For instance, if a comparison is made between two road projects costing $1 million—one a gravel road with an economic life of 5 years and another an asphalt road with an economic life of 20 years—then the gravel road project should be repeated 4 times to make the life of the two projects equal.

However, a small one-year or two-year difference in projects with long lives of 25 to 30 years could be ignored as the *NPV* of output realized after 25 years will be relatively small.

Economic Internal Rate of Return

The economic internal rate of return (*EIRR*) is commonly used by international institutions and bilateral agencies in determining project viability. This is primarily because the investment budget of a borrowing country is very often not known, nor are there many readily appraised projects to choose from. More importantly, it may be difficult to determine the *OCC* which could be the basis for estimating the *NPV* of a project. The advantage of the *EIRR* is that it can be calculated based on project data alone, and a decision on project viability can be made without knowing precisely the *OCC* of a specific country.

The *NPV* and *EIRR* methods are, however, closely related. The NPV method, through the use of the discount rate, can estimate the present value of all benefits and costs. The difference between benefits and costs represents the *NPV*. In the *EIRR* calculation, the procedure is reversed: The *NPV* is fixed at zero and, through a process of trial and error, the discount rate which will make equal the present value of the flows of costs and benefits over the life of the project is determined. The *EIRR* of a project is obtained by the solution of i in the following equation, a modified version of equation 5 which estimates the *NPV*:

$$\sum_{t=0}^{n} \frac{B_t - C_t}{(1 + i)^t} = 0 \qquad (6)$$

where B_t represents the expected benefits in year t of a project's life and C_t represents the costs expected to be incurred in year t. The length of life of the project is denoted by n.

The *EIRR* provides a simple and readily comparable measure of the profitability of a project; the higher the *EIRR*, the better a project and vice-versa. The *EIRR* is similar to the benefit-cost ratio, discussed below, as both ignore the size of a project. For decision makers both in developing countries as well as in donor agencies and institutions, it is much easier to understand and appreciate the statement that a project yields a return of i, than the statement that the *NPV* is $10,000 at a discount rate of r. Furthermore, it is much easier to compare the two *EIRR*s at their appraisal and evaluation stages than the two *NPV*s.

Given the data for equation (6) mentioned above, it is fairly easy to work out the *EIRR* of a project through the use of a calculator or a personal computer. Table 9.2 shows a simple example of how to calculate the EIRR. The cash flow data (*B–C*) are the same as given in table 9.1.

The *NPV* based on a 12 per cent discount factor added up to ₱3.845 million. The next calculation made with a discount factor of 20 per cent yields a negative *NPV* of ₱0.344 million. This would indicate that the *EIRR* of the project, where *NPV* = 0, is between 12 and 20 per cent and closer to 20 than to 12. To arrive at the discount rate which will yield *NPV* = 0, we divide ₱3.845 million by the difference between the two *NPV*s, i.e., ₱4.189 [3.845 – (–0.344)]. The resultant figure 0.92 represents the distance we have to move between the two discount rates (8 per cent points) to arrive at the correct discount rate. This gives a discount rate of a little over 19 per cent [12 + (0.92 x 8)]. Using the discount rate of 19 per cent, the *NPV* works out to be ₱0.054 million, which approximates an *NPV* of zero. The *EIRR* of the project is, therefore, 19 per cent.

TABLE 9.2
Example of Economic Internal Rate of Return
(Million Pesos)

Year	*Cash Flow (B–C)*	*Present Value at 12 Per Cent Discount Factor*	*Present Value at 20 Per Cent Discount Factor*	*Present Value at 19 Per Cent Discount Factor*
0	–9.000	–9.000	–9.000	–9.000
1	–2.000	–1.786	–1.666	–1.680
2	1.000	0.797	0.694	0.706
3	4.000	2.848	2.316	2.372
4	4.000	2.544	1.928	1.992
5	4.000	2.268	1.608	1.672
6	4.000	2.028	1.340	1.404
7	4.000	1.808	1.116	1.180
8	4.000	1.616	0.932	0.992
9	2.000	0.722	0.388	0.416
Net Present Value		3.845	–0.344	0.054

Benefit-Cost Ratio

The benefit-cost ratio (BC ratio) is also commonly used in ranking projects and determining their comparative economic viability. The BC ratio makes estimation simple and project comparison easy. The BC ratio is calculated by dividing the sum of the present value of all benefits by the discounted sum of all costs. The discount rate used is equal to the OCC. All projects which have a BC ratio higher than one are considered acceptable. The higher the ratio in excess of one, the more acceptable the project.

This approach, however, suffers from certain serious shortcomings. First, it gives incorrect and misleading ranking when sizes of projects vary. Smaller projects with a higher BC ratio are favored over larger projects which may have a lower BC ratio but have higher net present values and, therefore, are more beneficial to the economy. This can be demonstrated through the example given in table 9.3.

The table above shows that based on the BC ratio, project X appears superior to project Y. But if the choice is to implement either of the two projects, it would be better to implement project Y, because it adds to net national income a much larger absolute amount than project X (600 as against 200). The BC ratio, however, yields correct results only if both projects are of equal size.

TABLE 9.3
Example of Benefit-Cost Ratio
(Present Value Discounted at 12 Per Cent)

Item	*Project X*	*Project Y*
1. Investment Cost	500	2,000
2. Benefits (2a – 2b)	700	2,600
2a. Gross Benefits	1,000	3,500
2b. Operating Cost	300	900
3. Net Present Value (2 – 1)	200	600
4. Benefit-Cost Ratio		
4a. Alternative I [2a ÷ (1 + 2b)]	1.25	1.21
4b. Alternative II (2 ÷ 1)	1.40	1.30

The BC ratio has another serious limitation. The ratio is not unique; it varies with the way costs and benefits are defined. For example (in table 9.3), if benefits are defined as gross benefits while the cost includes both investment and operating cost, then the BC ratio for project X works out to 1.25. If, on the other hand, benefits are defined as net benefits excluding operating cost, then the BC ratio would be 1.40. Since there is no standard procedure for defining costs and benefits, the use of the BC ratio could lead to incorrect comparison and wrong decisions. In view of these shortcomings, the use of the BC ratio is not recommended for project selection.

Pay-Out Period

In its simple and often used form, the pay-out or payback period measures the number of years it would take for the net benefits (positive cash flow) to pay back the investment. If the pay-out period is longer than the one arbitrarily chosen, say, five or six years, the investment is rejected; if the period is shorter, it is accepted. In its more sophisticated form, this method compares the net present value of benefits with the net present value of costs for a given period. If the benefits exceed the costs, the project is accepted; otherwise, it is rejected.

The basic assumption made in the pay-out period method is that benefits beyond a certain period are highly uncertain and should be ignored. While it is undoubtedly true that predicting the future is always fraught with uncertainties which increase with successive years, it would be wrong to use this as a basis for arguing that all project benefits beyond a given period should be ignored. This approach may have some validity in respect of small-scale private-sector projects whose economic life is, in any case, so short that benefits beyond five to six years are not likely to be very significant. But in the case of public sector projects where investment cost itself is spread over several years and it takes many years for the full benefits to be realized, this approach is inappropriate and misleading. If this approach were adopted, many of the worthwhile long-gestation projects in the public sector will not be taken up.

If the pay-out period method is used as a basis for project selection, there is a risk that a wrong choice of projects will be made and less economically viable projects will get selected. This point is illustrated through an example given in table 9.4.

TABLE 9.4
Example of Pay-Out Period

	Cash Flow		*Discount Factor*	*Discounted Present Value*	
Year	*Project X*	*Project Y*	*at 12 Per Cent*	*Project X*	*Project Y*
0	–2,000	–2,000	1.000	–2,000	–2,000
1	600	250	0.893	536	223
2	600	300	0.797	478	239
3	550	400	0.712	392	285
4	500	450	0.636	318	286
5	450	500	0.567	255	283
Total (1 to 5)	2,700	1,900		1,979	1,316
6	400	600	0.507	203	304
7	300	700	0.452	136	316
8	200	800	0.404	81	323
9	100	800	0.361	36	289
10	100	800	0.322	32	258
Total (1 to 10)	3,800	5,600		2,467	2,806

Table 9.4 gives the example of two projects, X and Y, which have the same capital cost in the base year and whose benefits flow over a period of 10 years. In project X, the benefits are high in the initial years but steadily decline over time. The reverse is the case in project Y, where the benefits build up over the years, being the highest at the end of the period. If one used an arbitrary period of five years as a basis of selection, then project X appears superior to project Y, but if the entire flow of benefits over the economic life of 10 years is taken into account, project Y turns out to be economically more profitable than project X. This is true whether one used the discounted present value or an undiscounted cash flow approach.

The major objection against the pay-out approach is that it places a premium on projects which have short lives and whose benefits are generated quickly. On the other hand, it penalizes projects whose benefits extend over many years and build up slowly. Since this textbook deals primarily with public sector projects which fall mostly in the second category, it is recommended that the pay-out period not be used in deciding the choice of public sector investment projects.

Conclusion

Project viability is determined by comparing the streams of net benefits and costs over the life of a project. Since the flows of costs and benefits take place over different time periods, they cannot be directly compared. The principal criteria used for making a meaningful comparison are the NPV and the EIRR. Both require the determination of a suitable discount rate which represents the OCC to the economy. Under the NPV method, the discount rate is used to determine the net present value of both the cost and benefit streams. If benefits exceed costs, the project is considered viable. The EIRR constitutes the discount rate which equates the present value of the streams of benefits and costs. If the EIRR exceeds or is equal to the OCC, the project is accepted and if it is below the OCC, it is rejected.

Both criteria are important in project selection. The EIRR is more commonly used by international lending institutions which are often requested by borrowing countries to finance specific projects. The EIRR provides a single measure of the return of the project on which a decision can be taken about its acceptability. The higher the EIRR, the better the project. The NPV criterion is particularly valuable in making a selection of projects to fit in the available resources. In this case, the projects which maximize the NPV should be chosen. The EIRR method cannot be used for this purpose.

In several instances, in addition to the NPV or the EIRR, the feasibility reports also provide information on the number of workers employed and the net amount of foreign exchange earned or saved by the project. An impression is given as if these contributions, beneficial or otherwise, are in addition to the benefits measured by the NPV or the EIRR of the project. This is misleading and incorrect because the wage rate, the number of workers, and the foreign exchange impact are already captured in the economic benefit and cost flows of the project.

This point will be obvious if we look at what the EIRR analysis attempts to accomplish. In this analysis, economic values of components of costs and benefits are already included and the EIRR is measured by the rate which equates the two flows. Tests similar to that of the EIRR can be derived for labor or foreign exchange and the critical points of acceptance or rejection will then become the shadow wage rate or shadow exchange rate. All such tests are equivalent as long as inputs and outputs remain the same and no new information is fed into the cost or benefit streams of the project. The EIRR (or the NPV) should, therefore, be regarded as an all-inclusive measure of the viability of the project. The other aspects, if mentioned, should be treated as part of the overall return of the project.

CHAPTER 10

Sensitivity Analysis and Project Risks

Introduction

The NPV and the EIRR discussed in chapter 9 are estimated by using the most probable values of the key parameters in the cost and benefit streams of a project. However, the values of these parameters can change over the life of a project for reasons which could be difficult to predict at the time of appraisal. For instance, the quantities and values of inputs and outputs may turn out to be considerably different from those anticipated in the appraisal. Since the appraisal of a project would become a highly complex and detailed exercise if its viability is tested with reference to a combination of different values of inputs and outputs, a simpler approach using sensitivity analysis is generally adopted to test the effects of possible changes in the value of key parameters on the viability of the project.

Sensitivity analysis is also important in determining the risks inherent in a project. The risks, however, vary greatly with the nature of the project. If a high degree of risk is apparent from the results of the sensitivity analysis, the project's appraisal report should incorporate measures to minimize such risks. In social sectors where sensitivity analysis is not feasible, the project risks should be related to the socioeconomic objectives sought to be achieved by the project. This aspect should be highlighted in the project's appraisal report so that policy makers can take an informed decision whether to proceed with the project in the light of the risks identified.

This chapter discusses the purpose and method of sensitivity analysis and the assessment of project risks. It may be mentioned that since the sensitivity analysis shows the impact of a change in a variable on the viability of a project, the results are better handled through the use of the EIRR. For instance, a reduction in project output will reduce the EIRR compared with the base case EIRR. The two EIRRs can be readily compared and a decision taken about the degree of risk involved. Such a comparison becomes more complicated when the NPV is used, where the impact is on the absolute value of NPV. Consequently, the discussion in this chapter is confined mainly to the impact of sensitivity analysis on the EIRR of a project.

Sensitivity Analysis

Purpose and Method

Sensitivity analysis is carried out to determine how possible changes in underlying assumptions affect a project's viability. While the EIRR is determined by the projected magnitudes of the costs and benefits over the life of a project, sensitivity analysis shows how changes in key variables affect the EIRR of a project. Sensitivity analysis essentially involves four steps: First, the key variables relevant to the project are determined. Second, the changes in the quantity and/or value of each variable most likely to occur are estimated. Third, the effect of each change on the cost and benefit stream and the calculation of the EIRRs resulting from these changes are determined. And fourth, the results obtained are interpreted and their implications explained.

It is neither feasible nor necessary to examine the effects of all possible changes in each of the variables on the EIRR. The main variables most likely to be affected would vary with projects and these should be identified during appraisal in consultation with engineers and the technical staff. Some variables that affect practically all projects and should always be included in the sensitivity analysis are: (1) prices and quantities of outputs; (2) prices and quantities of inputs, both for construction and operation of project facilities; (3) the level of capacity utilization of project facilities; and (4) the time taken to commission the project to achieve target capacity operation.

The prices of project outputs influence the size of the benefit stream, while the prices of project inputs affect the cost stream.[1] These changes can greatly affect the viability of a project. The prices of some commodities, particularly primary commodities, often fluctuate widely and changes are difficult to predict. However, as mentioned in chapter 7, the alternative prices of major inputs and outputs that might prevail should be estimated and their impact on the project EIRR shown through sensitivity analysis.

The quantities of goods and services required for constructing project facilities may be greater than originally estimated. That is possible more for certain types of projects, e.g., multipurpose dams and port projects. A sensitivity analysis should be undertaken to determine the impact of the increase in cost on the EIRR of the project.

The quantities of project outputs may be less than anticipated for a variety of reasons. This may, for instance, be due to an inadequate supply of spare parts or electricity in an industrial project or an insufficient quantity of irrigation water or fertilizers in an agricultural project. Further, the quantity of project outputs produced per unit of inputs may not attain the rate assumed in the calculation of the base case EIRR because of an inefficient use of inputs, a shortage of skilled labor, difficulties in adjusting to new technology, or inefficient management. The possible impact of these factors on production should be considered in a sensitivity analysis.

The level of capacity utilization of the project facilities may also be lower than projected, thereby affecting both the benefit and cost streams of the project. The extent of capacity underutilization in power, transportation and industrial projects can be expressed as a percentage of the level of utilization assumed in calculating the base case EIRR. In irrigation projects, underutilization, as indicated by the expected reduction in cropping intensity, can be expressed as a percentage of that assumed in the project. A lower-than-optimum level of capacity utilization reduces output and hence project benefits, but it also reduces operating costs. The effects of capacity underutilization should, therefore, be expressed in terms of the net effect on the project.

The time required to commission project facilities and to achieve the target level of capacity utilization may be longer than that envisaged in the project's appraisal report. The delay in completing project facilities will affect both the cost and benefit streams of the EIRR. It is also possible that even if the project facilities are completed on schedule, there may still be a delay in attaining the target level of output. Here, both the cost and benefit stream will again be affected. The effects of these delays should therefore be expressed in terms of the net effect.

The overall impact of the change in a variable on the EIRR of a project will depend upon the anticipated per cent change in the variable and its importance in the overall cost or benefit stream of the project. For instance, if the project has a single output and there is a risk of it being too low against what is used in the base case during appraisal, the EIRR could be affected greatly in the sensitivity analysis. If, on the other hand, only a few minor items on the cost side are expected to be affected, and major items represent firm estimates, the overall impact on the EIRR under sensitivity analysis would be relatively small.

Checklist for Sensitivity Analysis

In carrying out the sensitivity analysis, the following aspects should be taken into account:

- The key variables should be examined and possible changes in them from the levels used in the base case EIRR should be clearly stated.[2] The basis for assuming these changes should be explained.
- Changes in the prices or quantities of inputs and outputs that will affect either benefits or costs should be expressed in terms of the resulting per cent change in the cost or benefit streams.
- The range of EIRRs implied by the assumed changes in the benefit or cost stream should be calculated and presented.
- Since several unfavorable changes may occur together, the results of simultaneous adverse changes should also be presented. In particular, the combination of changes which have common causes, or which tend to occur together should be tested.
- In cases where EIRRs are calculated for individual subprojects, a sensitivity test should be applied to each subproject.
- The sensitivity indicator (SI) which shows the sensitivity of the EIRR to changes in the variable tested should also be presented for all the variables tested (see also table 10.3).

- If sensitivity analysis reveals that the viability of a project is highly sensitive to changes in any key variable, its base case estimate should be carefully examined. If necessary, safeguards to alleviate or obviate such changes should be incorporated in the project's appraisal report.

A Numerical Example

To illustrate the use of sensitivity analysis, an example is given of a project involving the construction of a mill producing two products: A and B. The share of product A in total revenue is 60 per cent and product B, 40 per cent. The mill is to be built in four years and is to be operated for 20 years, after which it will have only scrap value.

Key Variables

Table 10.1 gives the quantities and values of the key variables used in the calculation of the base case EIRR and those chosen for sensitivity analysis. The changes in variables that are tested should represent the most likely occurrences and should be based on the experiences of similar projects in the past. Examples of the reasons for the lower figures used in table 10.1 for sensitivity analysis are explained below.

TABLE 10.1
Quantities and Values of Key Variables Used in Base Case and in Sensitivity Analysis

Variables	*Base Case*	*Sensitivity Analysis*	*Percentage Change*
Price of A	$10/kg	$9/kg	–10
Price of B	$33.33/kg	$30/kg	–10
Productivity of A	10 mt/day	9 mt/day	–10
Productivity of B	2 mt/day	1.6 mt/day	–20
Cost of Raw Materials	$500/mt	$600/mt	+20
Cost of Fuel	$20/barrel	$28/barrel	+40
Capacity Utilization	300 days	270 days	–10
Construction Costs			+10
Commissioning of Projects	December 1994	December 1995	Delay of 1 year

Prices of both products A and B have been estimated in the base case, taking into account the past trend and the likely increases in world supply and demand. There is, however, a possibility that the world prices may show a decline, especially due to reduced demand in industrial countries. A worst-case scenario of a 10 per cent decline in the prices of both commodities is assumed for sensitivity analysis.

The productivity of A and B in the base case is derived from the most efficient plants operating in the country. The project entity has experience in producing A, but productivity was low due to old equipment. The labor force employed has no experience in producing B, which is a new product for the managers of the project. To examine the sensitivity of project benefits, a productivity decline of 10 per cent for A and 20 per cent for B is tested.

On the cost side, prices of raw materials are based on recent world prices which have been low due to an abundant supply from major producers. This favorable situation may not continue. An increase in the cost of raw materials of 20 per cent, therefore, is tested.

The price of fuel oil has also been based on a recent price trend which has been low in real terms. Considering the low known international reserves in relation to the expected growth in world demand, an increase of 40 per cent in fuel oil costs above the base case level is chosen for testing.

Other production costs, including wages which account for 40 per cent of the total, are not expected to show any significant change.

Due to electricity interruptions in recent years, plants have not been able to operate at full capacity in the project area. A 10 per cent reduction in capacity operation is, therefore, tested.

The prices of machinery, equipment, and raw materials used in estimating project cost appear appropriate. However, there may be some underestimation in the physical quantities required in completing the project. The effect of a 10 per cent increase in construction costs is, therefore, tested.

The project is scheduled to be completed by December 1994. The period allowed for project completion should normally be adequate, but considering the lack of experience of the project authority in implementing similar projects, the effect of a delay of one year is tested in the sensitivity analysis.

Results of Sensitivity Tests

The results of the sensitivity analysis are presented in tables 10.2 and 10.3. Table 10.2 shows the effects of the changes in key variables on the benefit and operating cost streams. Table 10.3 shows the range of EIRRs calculated under various assumptions and the sensitivity indicator for each. The sensitivity indicator represents the per cent change in EIRR as a result of a per cent change in the variable tested. The higher the sensitivity indicator, the greater the potential risk.

Table 10.3 shows that the EIRR is most sensitive to changes in the benefit stream arising from changes in prices and productivity. The project is also sensitive to reductions in capacity utilization. It is, therefore, important that the assumptions made in these areas are realistic. At the same time, safeguards should be built into the project to ensure that the risks of reduced productivity and capacity utilization are minimized.

TABLE 10.2
Effects of Changes in the Values of Key Variables in Operating Costs and Benefits

	Changes in the Values of Key Variables			*Percentage Shares of Total*	*Changes in Total*
	From	*To*	*(%)*	*(%)*	*(%)*
A. Operating Costs[a]					
1. Fuel cost	$20/barrel	$28/barrel	+40	10	+4
2. Raw materials	$500/ton	$600/ton	+20	50	+10
3. Combination of (1) and (2)				60[a]	+9
B. Benefits					
1. Price of A	$10/kg	$9/kg	−10	60	−6
2. Productivity of A	10 tons/day	9.0 tons/day	−10	60	−6
3. Price of B	$33.33/kg	$30/kg	−10	40	−4
4. Productivity of B	2 tons/day	1.6 tons/day	−20	40	−8
5. Combination of (1) and (3)			−10	100	10
6. Combination of (2) and (4)			-14	100	14

[a]The remaining 40 per cent comprises labor and other costs.

TABLE 10.3
Sensitivity Analysis

	Change (%)	*EIRR (%)*	*Sensitivity Indicator*[a]
Base Case	–	22.01	–
1. Investment Costs	+10	20.36	0.75
2. Operating Costs	+15	20.66	0.41
3. Output Prices	–10	19.24	1.26
4. Productivity	–15	18.52	1.06
5. Capacity Utilization	–10	19.81	1.00
6. Delay in Construction	one year	20.21	–
7. Combined Unfavorable Assumptions			
(a) Combination of (1), (2), and (6)		17.63	
(b) Combination of (1), (2), and (3)		16.32	
(c) Combination of (4), (5), and (6)		15.33	

[a]The sensitivity indicator (SI) is calculated as follows: SI = percentage change in EIRR/percentage change in the variable tested. For example, when SI for output prices is 1.26, it means that a 10 per cent increase in operating costs leads to 12.6 per cent decrease in the EIRR, i.e., from 22.01 per cent to 19.24 per cent. The sign of the indicator (positive or negative) need not be shown.

The project is much less sensitive to increases in investment and operating costs. This means that even a significant increase in these costs will not seriously erode the viability of the project.

To determine the impact of a simultaneous adverse change in certain variables, three combinations are tested in table 10.3. The combinations tested are those involving key variables which are closely related and are likely to occur together. The first combination includes a delay in project completion, a cost overrun and an increase in operating costs. The second combination tests increases in both investment and operating costs and a decrease in the prices of outputs. The third combination tests a delay in construction, a decline in productivity, and a reduction in capacity utilization. These tests show that if any of the three combinations of adverse changes occur, the EIRR decreases substantially but the project remains viable since the EIRR remains above 12 per cent which is regarded as the OCC in the country.[3]

Project Risks

Projects with Quantified Benefits

Project analysis is not complete without a discussion of the risks involved in implementation. While the risks vary from project to project, the discussion will be primarily confined to those risks which entail major economic consequences. These should be identified from the sensitivity analysis discussed above and described in descending order of importance with regard to their impact on the EIRR.

Project risks are directly related to the magnitude of the sensitivity indicators of the key variables. In all cases where the variables tested are found to have sensitivities greater than one, they should be further disaggregated to identify the primary sources of sensitivity. The usefulness of sensitivity analysis basically lies in the isolation of individual parameters and in the identification of the sources of risk involved and, to the extent that they are due to factors which can be controlled or modified, to suggest appropriate measures. Even when the risks are beyond the control of the project authorities, sensitivity analysis makes the policymakers aware of the nature and extent of the risks involved, thereby enabling them to make an informed decision about the implementation of the project.

Particular attention should be paid to risks that would substantially reduce the project's EIRR or render the project uneconomic by reducing its EIRR below the opportunity cost of capital. In such a case, the base case EIRR and the sensitivity indicator are relevant. If the base case EIRR is high, the discussion of project risks should generally include risks to which the project is highly sensitive. For instance, the EIRR of most projects is highly sensitive to changes in project output. A discussion of safeguards employed to minimize the risk of output falling substantially below the level expected should, therefore, be included. For another example, apart from water availability, an irrigation project's output may depend upon the supply of other inputs, provision of extension services, effectiveness of water management by farmers' groups, and availability of adequate infrastructure and storage facilities. Measures taken to ensure adequate and timely availability of each should be explained.

Risks are obviously greater in projects for which the base case EIRR is only marginally higher than the OCC. The risk becomes even greater if the EIRR is highly sensitive to changes in key variables since even a small reduction in the EIRR would render the project unviable. Even when the EIRR is relatively insensitive to changes in key variables, a combination of adverse changes might easily affect the project's viability. Thus, in such cases, the remedial actions proposed or adopted to maintain the viability of the project need to be fully explained.

If major project inputs or outputs are traded internationally, a major risk may be future changes in prices. In such cases, a review of the world demand and supply forecasts for the goods in question and the basis of price determination should be included (see also chapter 7).

The risk analysis, described above, is based on single values of individual variables whose sensitivity is tested by assuming specific changes in them. In practice, these single values represent what the project analyst regards as the best estimate from a range of probable values. For the assessment of a range of actual possibilities and the probability of their occurrence, a more sophisticated risk analysis, known as probability analysis, can be conducted for certain projects either singly or in combination with the sensitivity analysis. Such an analysis is particularly essential for those projects—such as gas exploration—for which the uncertainty of the outcome is very high.

Probability analysis is undertaken by formulating and utilizing a complete judgment on the possible range of individual variables and the likelihood of each occurring within a range. The probability distribution of the risk is derived from subjective and intuitive judgments of the study team about the probabilities of the outcome. The probability distribution consists typically of the range and the distribution of values within it. The range is established by determining the highest and the lowest probable values of the variables considered. Starting normally with the best judgement value and making successive subjective estimates around it, the probable distribution of other values within the range limits can be formulated.

Once the probability distribution of significant variables is determined, a simulation program and an iterative process can determine the probability distribution curve of the project's decision variables and can calculate its mean and variation therefrom. From there, a range of possibilities for returns may emerge. For instance, instead of a single EIRR of, say, 15 per cent, it may be possible to conclude that the probability of an

EIRR of 12 per cent is 80 per cent and the probability of an EIRR of 10 per cent is 95 per cent. Due to the complex and difficult nature of the work involved, the probability analysis of risks is normally undertaken only for projects carrying a high degree of risk, or for large projects where miscalculations could lead to a major loss to the economy. For such projects, the nature of the risks involved and the measures proposed to minimize them, together with the results of the sensitivity analysis should be discussed in the project's appraisal report.[4]

Projects with Unquantifiable Benefits

Projects in the social sector such as education, health, sanitation, and family planning have many unquantifiable benefits. It is not possible to quantify and value all the benefits and estimate the EIRR of such projects. Consequently, the risks cannot be measured by sensitivity analysis. In such cases, the relationship of project risks to project objectives should be explained. The eventualities that might impede the realization of the objectives should be discussed in relation to the project cost and output, as well as in relation to the socioeconomic objectives sought by the project.

In social projects, investment costs primarily relate to the construction of buildings and provisions for equipment and supplies. The risks on the cost side, therefore, relate to factors which could delay project implementation. These might include, for example, timely provision of financial resources, implementation capacity of the project authority, and availability of land. In such projects, the risks are far greater on the benefit side. For instance, in education projects school buildings and equipment are provided to help achieve a certain annual output of graduates with certain levels of skills. However, facilities alone may not ensure the achievement of the project's objectives. Their achievement may depend more upon the availability of trained teachers, the availability of sufficient funds for the recurring expenditures of the institutions, curriculum and admission standards, and student motivation.

While it is not possible to eliminate all risks, every effort should be made to minimize them. Major risks should be identified and explained and remedial measures should be proposed.

The major benefits of social projects relate to broad socioeconomic goals. For education projects, these may include an increase in income levels of trainees or higher levels of industrial and agricultural productivity. For family planning projects, the broad goals be an increase in the number of acceptors and a consequent reduction in population growth. The success of such projects thus depends not on the facilities provided but also on the existence or creation of favorable conditions as assumed in the project's appraisal report. For such projects, assumptions made regarding the relationship between the facilities provided and the project's long-term objectives should be clearly explained. The conditions or facilities necessary but external to the project should also be identified and necessary safeguards should be built into the project to ensure their existence. For such projects, this is one of the most important aspects to be discussed in the section dealing with project risks.

Endnotes

1. As noted earlier, constant prices are used in determining the economic viability of a project. Therefore, changes in prices mentioned in this chapter refer to relative price changes, unless otherwise noted (see also chapter 7).
2. Favorable changes would enhance the viability of the project. Since the purpose of sensitivity analysis is to show whether the project would remain viable if circumstances less favorable than those assumed in calculating the base case EIRR prevailed, only unfavorable changes are normally considered.
3. Throughout this chapter, sensitivity analysis is discussed with reference to the EIRR of a project. This analysis can be equally applied to projects in which the NPV is calculated. The main difference is that in the case of EIRR sensitivity analysis shows the extent of reduction resulting from a given change in a variable, while in the case of NPV, an estimate is made of the percentage change which will reduce the NPV to zero. The value of a variable at which the NPV becomes zero is called the "switching" or "cross-over" value. These switching values are helpful in providing a better understanding of the critical elements on which the outcome of a project depends so that the project authorities could devote further effort in firming up the estimates and narrow down the range of uncertainty.
4. For a more comprehensive discussion of the probability analysis, see Louis Y. Pouliquen, *Risk Analysis in Project Appraisal*, Occasional Paper Number 11 (Baltimore: The Johns Hopkins University Press, 1970).

CHAPTER 11

Project Benefit Monitoring and Evaluation

Introduction

Monitoring and evaluation (M&E) of projects is a relatively new discipline and interest in it emerged only during the past two decades. This interest arose primarily from the realization of the complex nature of the development process and the need to learn from experience. By now practically all international institutions and many of the bilateral donors incorporate M&E in their projects. The poor performance of many projects and the shortage of funds in developing countries have highlighted the importance of M&E. Many developing countries have also come to realize the importance of M&E for effective project management and have set up M&E units in individual ministries or national planning organizations.

The basic objective of M&E is to promote an efficient and effective implementation and operation of development projects. It helps minimize delays through the early detection of problems and by taking timely corrective action. M&E serves as a feedback system for problem solving during the implementation stage and helps in realizing the objectives of the project during its operation. M&E could even lead to a revision of the objectives themselves if, during project implementation, such objectives are found to be unrealistic or impractical.

Monitoring and evaluation should be undertaken both for the investment cost and benefits side of the project. In this chapter, however, attention is focused on the monitoring of benefits and on the economic evaluation of projects. This is not to suggest that efficient project implementation is any less important than the achievement of the benefits of a project. In fact, the two are interrelated. There are two reasons for focusing on monitoring of benefits. First, project implementation essentially involves the technical and engineering aspects which are outside the scope of this study. Second, the benefits of projects accrue over a long period and, therefore, the difference between anticipated and actual results tends to be far greater on the benefits side than on the cost side of the project.

Monitoring of Benefits

Monitoring and evaluation are related but are distinct activities. The monitoring of benefits is, however, essential for carrying out the evaluation of projects. Project benefits, in most cases, represent the difference between the output or status of beneficiaries with and without the implementation of the project. Such an estimate can be made only if the actual status of the project area is known before taking up the project; thereafter, the benefits of the project are assessed at regular intervals by estimating the difference in output in the with-project and without-project situations. To undertake such an estimate, a system of benefit monitoring must be built into the project at a fairly early stage in the project cycle.

A system of benefit monitoring is essential in all those cases in which economic activity exists in a given area without the project, and the benefits are represented by the incremental output resulting from its implementation. In an agricultural project, for instance, the benefits may be measured by the net value of additional output resulting from the project. This requires an estimation of the value of additional output in the project area. However, the full cost of all additional inputs required to achieve the additional output must be deducted from the value of this additional output. Similarly, in a road project, benefits will be represented mainly by the savings in operating cost of vehicles. This is measured by the difference in operating costs with and without road improvement.

Since output in the project area can change without the project, it is essential to carefully assess the output (benefits) that will be realized without the project. For instance, without the project, the condition of a road may deteriorate. In that case, the project should include among its benefits the savings in vehicle wear and tear. On the other hand, if without the project, some additional output of foodgrains would have taken place because of the recent introduction of new seed varieties in the project area, then that output should not be treated as a benefit of the project.

Estimating an area's future output without the project requires a careful identification of the various factors which have a bearing on the output and which assess their likely impact in the future. Ideally, this kind of assessment requires the selection of a "control group" or a "control area" comparable with that in the project area. Comparisons should also be made for changes occurring in both areas or groups. The purpose of these is to account for nonproject-induced changes such as weather, dietary habits, government policies, etc. Only when the effects of such exogenous factors are properly accounted for can the result of project-induced changes be properly evaluated.

In practice, it may not be possible to estimate all the changes induced by exogenous factors. For instance, it may not be easy to find a control group which will remain unchanged during the life of a project, but even if such a control group existed, it would be difficult to establish a causal relationship between project interventions and the variation in conditions of the project target groups/areas. In such cases, a rough estimate of the likely changes without the project may be attempted. In some instances, when in the judgement of the project analyst the changes without the project are expected to be minimal, a "before and after" comparison could be made. The implication of that comparison is that the entire increase in output is attributable to the project over the level which existed at the time of appraisal.

The with- and without-comparison requires the establishment of a baseline or bench mark before implementing the project. The baseline survey should be undertaken by an economist or statistician preferably at the feasibility study stage. The indicators to be included in assessing the baseline situation will depend upon the evaluation sought to be undertaken at a later stage. For example, in an agricultural project, the estimation of the direct benefits of the project would require an assessment of the level of output, the yield per hectare, the price of each crop, the use of inputs, and the income levels by size of holdings. To determine the wider impact of the project, it may be necessary to include household indicators on per capita food consumption, and the availability of health, sanitation, water supply, education facilities, credit institutions, etc.

The purpose of the baseline survey is to assess the economic situation prevailing in the project area before the start of the project. This will serve as a basis for comparison with the situation that will exist after the completion of the project. The baseline survey should, therefore, contain data which can be replicated in the future. Because of the scarcity of reliable data and the shortage of human and financial resources, the selection of indicators should be kept to the minimum consistent with the needs and objectives of the project.

Although indicators should be ultimately determined by the objectives of the project, there are certain rules of thumb which can be applied in selecting them. They should be:

Valid—i.e., they are supposed to measure what should be measured.

Reliable—i.e., verifiable or objective.

Relevant—i.e., relevant to the project objectives.

Sensitive—i.e., sensitive to the situation being observed.

Specific—i.e., specifically adapted to the particular project objectives.

Cost-effective—i.e., worth the time and money spent on them.

Timely—i.e., not very time-consuming in collection.

To assess the full impact of the project, the baseline survey should be repeated at least twice—once immediately after completion of the project and next when the full benefits of the project are expected to be realized. The second survey is particularly necessary for projects, especially in agricultural and infrastructure sectors where the build-up period of benefits tends to be quite long. For the sake of consistency and proper comparison, the coverage of subsequent surveys in terms of variables used and households covered should be the same as for the baseline survey.

Because of the high cost and the considerable amount of time involved, it is generally not practical to undertake comprehensive surveys of the project area or the target group. A properly designed sample survey can serve as an adequate basis for collecting necessary information, especially where quantification of data is involved. These sample surveys may be supplemented by in-depth studies with respect to those aspects for which simple indicators cannot be designed. These may, for instance, include community participation or quality of services in rural projects. Such in-depth studies should be tailored to meet the specific needs of each project and should be undertaken independently of the sample surveys, either as an exploratory study before the surveys or in pursuit of particular issues after the surveys.

Evaluation

Evaluation is a process for determining systematically the efficiency, effectiveness, and impact of a project or a program in the light of its stated objectives. The objectives themselves can be arranged in a hierarchy of two or more levels. For instance, the immediate objectives of an irrigation project may be the construction of a certain length of irrigation canals and roads of various types. The medium-term objective (or effect) may be one of increasing the production of various crops while the long-term objective (impact) may be one of improving the income or well-being of the farmers.

Stages of Evaluation

The purpose of evaluation is to improve the activities still in progress and to help management in planning, programming, and decision making, in the light of the experience gained. The evaluation of a project may be carried out (1) during implementation (ongoing evaluation), (2) at completion (terminal evaluation), and (3) some years after completion when the project's benefits and impact are fully realized. These three stages of evaluation are briefly discussed below.

Ongoing Evaluation

Ongoing evaluation involves an assessment of the implementation progress of the project and its continuing relevance, efficiency, and effectiveness. It also involves a review of the likely output and impact of a project. The review should normally be carried out midway during a project's implementation. Such an evaluation can help the decision makers by providing information about any adjustments that may be necessary in the scope, objectives, policies, or any other aspect of the project.

Ongoing evaluation helps examine whether the assumptions made regarding the different aspects of a project at the appraisal stage are still valid or whether changes are necessary to ensure that the overall objectives of a project are achieved. For instance, in some cases, project design or technology may be found inappropriate; in other cases unforeseen factors, external or internal, may require a revision of the initial assumptions. In other projects, the objectives themselves may have to be redefined in the light of the experience gained during implementation.

Terminal Evaluation

Terminal evaluation is undertaken shortly after the completion of a project—normally within one year. Such an evaluation is particularly useful for projects which have a short gestation period like manufacturing projects or agricultural projects involving credit or input supply. In such projects, there may be no need for an *ex-post* evaluation discussed below. Terminal evaluation is, however, necessary even for projects requiring *ex-post* evaluation. The status of benefits at the terminal period should be carefully reviewed for projects involving a long gestation period and corrective measures, where needed, should be proposed to expedite the realization of the full benefits of the project.

Ex-Post Evaluation

Ex-post evaluation is undertaken some years after project completion when the full benefits of the project have been realized. Since the gestation period of projects, within and among sectors, varies greatly, the timing of *ex-post* evaluation should be tailored to the specific needs of each project.

The purpose of terminal and *ex-post* evaluation is twofold: (1) to assess the achievement of the overall results of the project in terms of efficiency, output, and impact; and (2) to learn lessons for future planning through improved project formulation, appraisal, and implementation.

Evaluation should thus be seen as a learning process which provides insight on reasons for the success or failure of individual projects to national experts and policy makers, as well as to international organizations providing assistance for those projects. The purpose of such an evaluation is not to find fault or apportion blame if a project's objectives are not fully realized but to ensure that mistakes are minimized in the future. The very process of carrying out an evaluation can be as important as the conclusions drawn, since involvement in the process can often induce better understanding of the activities being evaluated and lead to a more constructive approach to their implementation.

Evaluation Methodology

Terminal and *ex-post* evaluations involve a reassessment of the technical, institutional, financial, and economic aspects of the project. The objective is to ascertain how the project has performed on the cost and benefit sides, in comparison with the assumptions made at the appraisal stage of the project. The deviation between the anticipated and the actual costs and benefits, and the factors responsible for them, should be adequately examined and explained. In doing so, a distinction should be made between the factors within the control of the project authority and those outside its control. For instance, the benefits from a foodgrains project may be lower due to a reduced availability of irrigation water and lower than anticipated world prices. The former could have been avoided through better planning, but the latter is due to factors over which the country has no control.

The primary focus of terminal and *ex-post* evaluation should be to re-estimate the economic and financial viability of the project and compare them with the estimates made during appraisal. This is because the economic and financial rates of return capture the effects of all factors both on the cost and benefits side of the project. While the EIRR is the main basis for determining the economic success of the project, the FIRR provides a measure of financial profitability which is essential for the continued financial health and sustainability of the project.

During the *ex-post* evaluation, an attempt should be made to determine the overall impact of the project in the area served by it. Such impact evaluation is particularly important for agricultural and rural development projects where the major objective is to increase rural income and reduce poverty. Impact evaluation could be carried out through in-depth studies of the economic situation in the project area initially at appraisal stage and subsequently prior to post evaluation. A better picture of the impact is provided if such a study is done for a group of projects implemented by one government ministry or department.

To determine the EIRR and FIRR of a project during evaluation, all benefits and costs should be expressed in constant prices of the year in which the project is completed. A complication arises where part of the cost is in foreign currency and part is in local currency, where price changes follow a different trend, and where exchange rates also vary over the years. Annex I gives a simple method of calculating the financial cost in constant prices of the year of project completion. The method for deriving the economic cost in constant prices from financial cost is the same as explained in the previous chapters.

Organization Structure for M&E

Both monitoring and evaluation are tools for analyzing data and producing reports for decision making. Monitoring analyses, supplemented by in-depth studies about the socioeconomic status of the beneficiaries before and after project implementation, provide the information base for terminal and ex-post evaluation. In this way, monitoring and evaluation are interrelated and form a unified system.

To ensure the studies' independence and objectivity, the benefit monitoring and evaluation unit should be independent of the project implementing authority. The unit should normally form part of the planning division of a ministry so that lessons learnt during M&E are incorporated in the formulation of future projects. In small countries, a central M&E unit could be set up within the planning agency. The same unit should normally undertake the M&E of all projects undertaken by a ministry as this will not only reduce the cost of M&E but also improve the quality of work done by the unit over time. The unit should also be made responsible for undertaking the sectoral evaluation of projects when a sufficient number of projects in one sector has been implemented. Such an evaluation should focus on common problems or issues encountered among projects because that could provide a useful basis for improving project planning in that sector.

The M&E unit should preferably be headed by a senior staff who should report directly to the head of the planning department of the ministry or the planning agency. This would lend proper status and ensure objectivity of results. The unit should be small and its professional staff should have expertise in economics and statistics. Depending on the need, additional staff may be hired on a temporary or on contract basis. For instance, if a baseline survey or some in-depth study is required, a research institute might be hired on a contract basis, or if it is to be undertaken by the unit itself, enumerators and other field staff may be hired temporarily.

Since the evaluation reports are likely to focus more on the shortcomings rather than on the achievements of the projects, the M&E staff is not likely to be popular with the senior staff involved in implementing the projects. The successful functioning of such a unit will, therefore, greatly depend upon the freedom and objectivity with which it is allowed to operate, and the confidence placed on it by the authority to whom it submits its reports.

ANNEX TO CHAPTER 11

Treatment of Exchange Rate and Price Changes in a Completed Project

The purpose of this note is to suggest a simple procedure for expressing in constant prices the cost of a project incurred over several years involving exchange rate and price changes. The objective is to first estimate the financial cost in constant prices which can then be expressed in economic prices.

Let us take an illustrative project started in 1987 and completed in 1990, having the cost data given in annex table 11.1.

To estimate the total financial cost of a project in constant prices of the year in which the project is completed, two steps are necessary. First, all domestic and foreign costs should be expressed in constant 1990 prices, using the domestic deflator for domestic costs and the dollar deflator for the costs incurred in foreign exchange. Second, all costs are expressed in local currency by using the exchange rate prevailing in 1990. Annex table 11.2 provides an example of the procedure to be followed based on the data given in annex table 11.1.

The foreign price index is normally based on the manufacturing unit value (MUV) Index prepared by the United Nations. This MUV is calculated by using the value of manufacturing exports of major industrialized countries as weights, and the exchange rate between $ and other currencies is used to arrive at the index expressed in dollar value. For the local currency cost, the wholesale price index (WPI) of the country concerned is used to determine the domestic deflator.

The method that may be used to estimate the economic cost from the financial cost given in annex table 11.2 is the same as described in the previous chapters.

ANNEX TABLE 11.1
Illustrative Data of a Completed Project

Item	*1987*	*1988*	*1989*	*1990*	*Total*
Foreign Cost: Current $ (million)	10	15	30	60	115
Foreign Price Index ($)	100	110	125	130	
Local Cost: Current ₱ (million)	50	60	80	150	340
Local Price Index	100	130	170	200	
Official Exchange Rate (Pesos per $)	14	17	20	21	

ANNEX TABLE 11.2
Method of Calculating Total Cost in Constant Local Currency (1990 = 100)

Item	*1987*	*1988*	*1989*	*1990*	*Total*
1. Foreign Cost: Current $ (million)	10	15	30	60	115
2. Foreign Price Index (1990 = 100)	77	85	96	100	
3. Foreign Cost in Constant 1990 $ (million) 3 = 1/2 x 100	13	17	31	60	121
4. Foreign Cost in 1990 ₱ (million) 4 = 3 x 21 ₱	273	357	651	1,260	2,541
5. Local Cost in Current Pesos (million)	50	60	80	150	340
6. Local Price Index (1990 = 100)	50	65	85	100	
7. Local Cost in Constant 1990 ₱ (million) 7 = 5/6 x 100	100	92	94	150	436
8. Total Cost in 1990 ₱ (million)	373	449	745	1,410	2,977

CHAPTER 12

Social Cost-Benefit Analysis

Rationale

The earlier chapters dealt with the various aspects of project analysis, with primary focus on economic efficiency. This analysis is also called traditional cost-benefit analysis (TCBA) to distinguish it from social cost-benefit analysis. In TCBA, the valuation of inputs and outputs is done in terms of the opportunity cost for the country. The objective of public investment here is to maximize economic growth.

The TCBA does not concern itself as to whom the benefits accrue and as to the purpose for which they are used. Implicit in this analysis is the assumption that an extra unit of consumption is as valuable as an extra unit of investment and that the marginal utility of an extra unit of consumption does not vary with income level. The economic efficiency analysis is not inconsistent with the broader social objectives. However, these objectives are sought to be realized through macroeconomic policies—such as fiscal and monetary policies—or specific programs designed to meet the needs of target groups or regions.

The social cost-benefit analysis (SCBA) was developed in the early 1970s with the support of international institutions, in recognition of the inadequacy of public sector resources to address persistent problems of poverty and inequality of income distribution in developing countries. Since the capacity of governments to increase savings and reduce poverty through macroeconomic policies was considered to be inadequate, project level analysis was seen as a means of redressing these problems. The SCBA is supposed to achieve these objectives.

Income distribution in SCBA has two dimensions: intratemporal and intertemporal. Intratemporal distribution focuses on income distribution between members of the present generation, while intertemporal distribution deals with income distribution across generations. The main concern of intratemporal income distribution is the problem of relative and absolute poverty and how public investment can be used to alleviate it through the promotion of the employment of lower income groups. The main issue in intertemporal distribution is the inadequacy of domestic savings. In this case, public sector investment is used as an instrument for promoting savings and increasing its availability to the public sector for promoting future socioeconomic development in the country.

The SCBA, therefore, differs from TCBA in that the former specifically incorporates in the valuation procedures the distribution consequences of using or producing goods and services. The economic efficiency objective under the TCBA is replaced by an explicit concern about who benefits from employment and income created by a project. Social accounting prices are used in SCBA compared with economic prices used in TCBA. The determination of the social accounting price is influenced by two basic assumptions: (1) extra consumption is worth more to the poor than to the rich and (2) an extra unit of consumption today may be worth less today than an extra unit of savings (investment).

The general effect of social accounting prices is that projects which generate greater consumption benefits to higher income groups are not favored. On the other hand, projects which mainly benefit the lower income groups of society are preferred as they show higher social returns because of the way the social accounting prices are defined. Consistent and comprehensive use of SCBA is expected to raise the level of investments over time, to change the sectoral allocation of public investment, to reduce income disparities, and to influence the choice of technology.

Determination of Social Accounting Prices

The starting point of SCBA is the determination of efficiency prices of both inputs and outputs. Thereafter, adjustments are made to reflect the effects on consumption and savings resulting from the use of inputs and outputs of the project. The consumption and savings effects generally work in opposite directions, and the net outcome depends upon the weight of these effects. Since the governments of developing countries often face difficulties in raising investment resources, a unit of income available to the government is considered to have greater value than if it was used for private consumption. At the same time, because of the declining marginal social value of consumption, an extra unit of consumption by lower income groups is given greater weight than that by higher income groups. The overall EIRR under SCBA will be greater than that calculated under TCBA if the distribution gains outweigh savings loss and vice-versa.

Let us illustrate the point with reference to the use of labor as an input in an agricultural project. Under the efficiency approach, labor cost is represented by the foregone output elsewhere in the economy. Let us assume that the employment of labor leads to a net increase in consumption. This represents a cost to the economy in terms of foregone savings. However, the additional employment may improve the income of the poorer sections of the society and this constitutes a benefit of the project. Both these factors combined with efficiency cost represent the social cost of an input to the project as shown below:

Net Social Cost of Input	=	Foregone Output at Efficiency Prices	+	Cost of Foregone Savings Because of Extra Consumption	–	Benefit Accruing From Extra Consumption

There is a certain level of private consumption at which an extra unit of such consumption is considered as valuable as an extra unit of investment. This level is termed as the critical consumption level or C_{cr}. For consumers below the C_{cr}, an extra unit of private consumption is worth more than an extra unit of public income. The reverse is the case for consumers above the C_{cr}. Using this reasoning, projects which benefit the poor become more viable under SCBA than under TCBA, while those benefiting richer groups become less viable.

A similar method is followed in determining the social benefits of the output of a project. The efficiency benefits of a project are added to the distributional effects on savings and consumption. If the additional output leads to greater consumption, the society will lose from foregone savings. However, to the extent that the benefits of additional consumption accrue to poorer sections of the society, they represent a gain from the project. The net distribution effect should be added to the efficiency benefit to determine the total social benefit of the project. Thus:

Social Benefit =	Efficiency Benefits	− Value of Foregone Savings Because of Extra Consumption	+ Benefits Accruing From Extra Consumption

Therefore, if the output of a project leads to additional consumption among people below C_{cr}, the social benefits would exceed efficiency benefits and the social rate of return will exceed the economic rate of return. However, if the consumption benefits accrue to higher income groups, the social rate of return will decline both because of foregone savings and a lower value attached to such consumption.

Valuation of Distributional Impact

As a starting point, it is necessary to clearly define the numeraire which in the efficiency analysis is defined as public income measured in foreign exchange. However, since savings and consumption under SCBA are to be valued differently, and consumption value depends on the income level of the individual, the definition of the numeraire must be sharpened to avoid an ambiguous measurement of benefits and costs. The numeraire is redefined as an uncommitted public expenditure measured in foreign exchange. Only uncommitted public expenditure is assumed to have a homogeneous unit value. All other expenditures can, in principle, have different values, and for each expenditure type, there is an accounting price that translates its value into units of numeraire.

Valuation of Income Distribution

The essential principles of valuation under SCBA can now be explained by making certain simplifying assumptions. Let us assume that project benefits are paid out either in the form of wages to workers or as surplus to the state, and that workers at all income levels have zero marginal propensity to save. At the margin, therefore, the government has to choose between holding on to the investible surplus and spending it on increasing private consumption both valued in terms of the numeraire. Let d_i be the weight attached to an extra unit of peso worth of consumption going to the ith group presently enjoying C_i level of consumption. Let $\bar{C}$ be the average (per capita) consumption level and let n be the rate at which utility increases relative to a unit rate of increase in consumption (or elasticity of marginal utility) at different income levels. Then

$$d_i = (\bar{C}/C_i)^n.$$

This means that as long as $n > 0$, an extra unit of consumption accruing to an income group whose present consumption is less than $\bar{C}$ will receive a weight greater than unity. The computed value will then depend upon (1) how poor the group i is relative to $\bar{C}$ level and (2) what value is selected for n.

Figure 12.1 helps clarify the principle. It is assumed that $n=1$ and that the average per capita consumption ($\bar{C}$) is ₱2,000 per annum. If the ith group has an income equal to this level, then the value of d_i is unity; where the current consumption level is lower (₱1,600), d_i is greater than unity (1.25). The poorer the ith group (C_i) relative to the reference income group ($\bar{C}$) and/or the larger the value of n, the greater will be the value of d_i.

The distributional weights thus greatly depend on the inequality of income distribution and the value of the elasticity coefficient (n). Table 12.1 gives a range of values derived from the distribution weights under different assumptions regarding n. For $n = 0$, an extra unit of consumption gets equal weight irrespective of the income level of the group to whom it accrues—an implicit assumption under the efficiency analysis. Since the value of distribution weights is an exponential function of n, the spread between the weights associated with the lowest and highest income groups greatly widens with the increase in the value of n. Determination of the correct value of n is thus of crucial importance in SCBA.

Valuation of Public Income

So far, the method of valuing the benefits accruing from extra consumption in the hands of different income groups has been examined. This, however, does not tell us anything on how a government values a marginal increas⌃ in its own investible sur-

FIGURE 12.1
Diminishing Marginal Utility of Extra Consumption

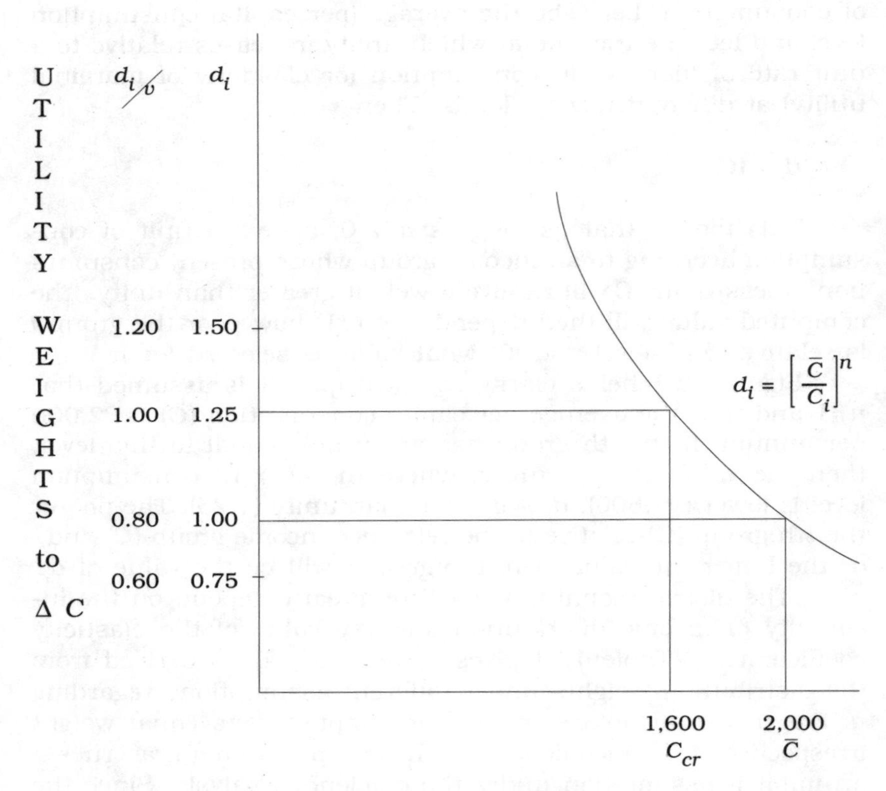

plus relative to a marginal increase in consumption to someone at the average level of consumption ($\bar{C}$). This ratio is conventionally denoted by the symbol v (for "value of public income"). If global savings is at a level where a government does not place a premium on raising savings relative to raising consumption of the average man (the optimal saving situation), the answer then is that $v = 1$ or $\bar{C} = C_{cr}$. If savings is sub-optimal, the value of public income must be greater than unity. The question is by how much.

TABLE 12.1
Values of Distribution Weights Using Alternative Values of *n*

Present Consumption Level (C) Peso Per Annum	$\bar{C}/C_i$	*Values of Elasticity Coefficient (n)*				
		0	*0.5*	*1*	*2*	*3*
1,000	2.00	1.00	1.41	2.00	4.00	8.00
$\bar{C}$ = 2,000	1.00	1.00	1.00	1.00	1.00	1.00
3,000	0.66	1.00	0.81	0.66	0.44	0.29
6,000	0.33	1.00	0.57	0.33	0.11	0.04
10,000	0.20	1.00	0.45	0.20	0.04	0.01

Two principal methods have been proposed. Little and Mirrlees (1974) suggested approaching this question directly by determining that level of consumption or "critical consumption level" (C_{cr}) at which the government is indifferent between spending more on consumption or on investments. In determining the C_{cr}, one would examine such things as minimum wage legislation, threshold level of taxation, welfare benefit entitlement, and subsidy policies. Using the example given in figure 12.1, let us assume that C_{cr} is found to be ₱1,600. The reference level then becomes ₱1,600 rather than ₱2,000. As a result, the group with per capita income of ₱1,600 is given a distribution (d) weight of unity and this alters the distribution weight of the other income groups. As will be seen in figure 12.1, this is equivalent to dividing the original d_i's by υ (or government savings standardized weights). The revised equation is given below:

$$(C_{cr}/C_i)^n = d_i/\upsilon.$$

Squire and Van der Tak or ST (1975) have suggested an indirect method which involves calculating a number of parameters and using C_{cr} as a crosscheck on the validity of the estimate. The ST formula for estimating υ is

$$\upsilon \ = \ \{(q - sq)/(i - sq)\}/CF_c$$

where

q = net return earned by a marginal unit of public investment,
s = marginal propensity to save (or reinvest) out of q,
i = the consumption rate of interest, and
CF_c = consumption conversion factor.

A major limitation of the above formula is that it assumes that the variables will remain constant over time. This assumption overestimates v because with the increase in investment, the divergence between q and i will decrease over time. Therefore, the value of v derived from the above equation should be treated as the upper limit of its true value.

The minimum estimate of v is derived by assuming that there is no reinvestment, or, in other words, there are no savings. The formula then becomes

$$v = (q/i)/CF_c.$$

However, even this formula could yield an overestimated v if the influence of the upward bias of the assumed constancy of q and i over time is greater than that of the downward bias of the elimination of reinvestment.

Social Wage Rate

In chapter 6, the economic or shadow wage rate (*EWR*) was defined as the marginal output of labor displaced from one sector of the economy by a new project in another sector of the economy.

The social wage rate (*SWR*) is different from the *EWR* because the employment of an extra worker commits the government to provide extra consumption or the difference between the wage rate and the net foregone output. Extra consumption now is clearly a benefit from the point of view of the worker, but it also represents a cost in the sense that it constitutes a claim on the limited investment resources and hence implies sacrificing future consumption.

The difference between the value of an additional worker's present consumption and the value of future foregone consumption constitutes the social cost element. This cost element must be added to the *EWR* to arrive at the *SWR* of hiring an extra worker:

$$SWR = m_i CF_m + \{(W_i - m_i) - (W_i - m_i)d_i/v\}\ CF_c$$

The first term simply represents the *EWR* expressed in border prices. The focus is on the social cost component contained in parenthesis and which contains two elements $(W_i - m_i)$ and $(W_i - m_i)di/v$. The former represents the extra consumption cost which is borne by the society as a result of hiring the worker. The latter expression represents the social assessment of the benefit to the worker which results from his receiving $(W_i - m_i)$ extra consumption, i.e., $(W_i - m_i)$ must be multiplied by a standardized distribution weight appropriate to the *i*th worker's consumption group (d_i/v).

From figure 12.1 it will be seen that if the worker belongs to that group for which extra consumption is valued on a par with extra investment $(d_i/v = 1)$, the whole expression in parenthesis drops out and the *SWR* becomes equal to the *EWR*. On the other hand, if the worker belongs to the poorer (richer) consumption group, the extra cost element will be negative (positive).

The social cost element in the *SWR* can be illustrated with the help of an arithmetical example. Let us assume that the worker's wage (W_i) under the project is ₱40, the foregone output (m_i) is ₱15, and CF_m and CF_c are both 0.80. Let us further assume that workers are drawn from the poor group whose standardized distribution weight is 1.6. On the basis of these assumptions, $EWR = m_i \cdot CF_m = 15(0.80) = 12$. The extra social cost element is

$$\{(W_i - m_i) - (W_i - m_i)d_i/v\} = \{(40-15)-(40-15)1.6\} = -15.$$

Applying the conversion factor of 0.80, the foreign exchange value of the extra social element is minus ₱12. Thus, the *SWR* = (*EWR* + social cost) = 12 + (−12) = 0.

Thus, while the efficiency cost of an additional worker expressed in foreign exchange is ₱12, the social cost of that worker is zero. If the distribution weight was higher than 1.6, the social cost would be negative, while, if the distribution weight was less than 1 (i.e., highly paid workers were employed), the *SWR* would exceed the *EWR*.

Value of Private Savings

So far it has been assumed that all factor payments made by the government lead to additional consumption. In practice, only a portion will be used for consumption, another portion will be taxed directly, and the balance will be used for savings and/or investments. Therefore, the cost and benefits of the transfer of resources from the public to the private sector depend not only on the foreign exchange cost of consumption and its social benefit but also on the social cost and benefit of the portion that is saved or invested.

Direct taxes do not involve any transfer of resources from the public to the private sector since the disposable income of that sector is effectively reduced by the amount of the tax. Direct taxes should, therefore, be netted out in determining the social benefit and cost of private income. Some part of private saving will be used for investment. Like government investment, private savings will have both a social cost and social benefit, and its overall social impact should be evaluated on the same lines as that of public investment.

A part of private savings may take the form of loans to the public sector. Since the loans involve obligations of a repayment of principal and interest, these should not be treated as equivalent to tax payment by the private sector.

Implications for Project Analysis

The use of the SCBA is likely to alter greatly the project design and shift sectoral priorities vis-a-vis those under efficiency analysis. This would be particularly so if the income distribution is highly uneven and the government attaches priority to reducing income disparities among its populace. This will lead to a choice of projects in which the labor content is high and the beneficiaries are low-income earners like landless laborers to whom high standardized distribution weights (d_i/v) are attached. In such cases, the *SWR* would be very low and even negative. Hence, such a situation will greatly raise the benefits and reduce the social cost of a project, thereby increasing its social viability compared with other projects under efficiency valuation.

The SCBA would also favor projects which generate or save public revenues. Projects which contribute a limited amount of savings to the government and whose benefits accrue to high-income groups will fall at the bottom of the priority list. In such projects, social viability will be lower than economic viability. From both inter-temporal and intra-temporal distribution considerations, such projects will be considered unsuitable for implementation by the government.

As shown in table 12.1, the distribution weights can greatly vary depending on the income group to whom the benefit accrues and the value of the elasticity coefficient. At a high value of elasticity coefficient (*n*), the social cost of hiring a worker can become negative. Consequently, some of the projects that are considered unsuitable for investment under efficiency tests may be considered viable under the SCBA. The implementation of a large number of projects of such a nature may, while improving the income of the poor people, prove burdensome to the exchequer as they would greatly reduce resources available for future investment. The risk is especially great if a high premium is not attached to government income.

It is, therefore, important that the values of the parameters are carefully chosen and the SCBA analysis is applied to the early stage of the project cycle and to all projects in the different sectors of the economy. The use of the efficiency criterion in one set of projects and the social criterion in the other set of projects could lead to an overall investment program which is both economically and socially inferior. If the use of SCBA is accepted, its consistent application across all sectors of the economy becomes essential.

Practical Problems in Adopting SCBA

This chapter described briefly and in simplified form the use of SCBA and how it differs from TCBA. The SCBA has a great deal of intellectual appeal because it simultaneously addresses the twin problems of domestic resource constraint and poverty which have been experienced by many developing countries in the selection of public sector projects.

However, despite the methodological advances made in the SCBA in the 1970s, its actual use in project analysis has been extremely limited both in developing countries and by external aid agencies. International institutions like the World Bank, UNIDO, and OECD which sponsored a great deal of research and studies on the SCBA have virtually made no use of this approach in project appraisal.

The reasons for the wide gap between theoretical studies and their empirical use are not difficult to understand. Practical problems and complexities faced by planning agencies in developing countries make it difficult to determine the key variables essential for undertaking the SCBA. Many planning agencies struggle to maintain rudimentary project appraisal practices due to data and staff constraints. Data needs for estimating distributional impact are enormous with macro, sectoral, and micro variables interacting at every stage. Furthermore, the determination of major variables require value judgement at the highest political level which is rarely exercised.

The complexities involved in the SCBA can be illustrated with reference to the distributional weights and savings premium mentioned above. The estimation of distribution weights at any point requires the availability of reliable data on income distribution by regions, by rural and urban areas and for the entire population. Such data is not readily available for many countries. Distribution weights can be applied only if consumption changes resulting from the project and the income group or the beneficiaries to whom they accrue can be reasonably determined. More importantly, the distribution weights depend on the estimated value of elasticity of marginal utility and the determination of reference income with which other incomes should be compared. A sound judgement on these variables is vital because minor changes could lead to a considerable change in distributional weights.

Concerning the value of n, it has been suggested that countries which express mild interest in redistribution could have a low value of less than 1. Such a recommendation may sound paradoxical if one makes the reasonable assumption that countries which show little interest in redistribution are likely to be the most in need of it.

The most difficult part of estimation concerns the changes in distribution weights over the life of the project. Since most of the projects have a life ranging between 15 and 50 years, the task of estimating distribution weights for each year is extremely complex and beyond the expertise and resources of most planning organizations. It is sometimes suggested that income distribution weights may be assumed to remain unchanged in future years. This is, however, inconsistent with the key objective of SCBA which is to reduce income disparities among population over time.

Similar problems arise in determining the social value of public income or budgetary premium (v). The critical factors in determining the value of v are the marginal product of capital (q) which is defined as net of the cost of maintaining capital intact; the marginal propensity to save (s) or consume ($1-s$); the rate at which the reinvested part saved continues to produce added flow of consumption benefits in the future $(1-s)q$; and the consumption rate of interest ($CRI = i$) or the rate at which consumption benefits are discounted to get their present value. Furthermore, assumptions have to be made about changes in the value of these variables over the life of the project. This list shows that the practical problems involved in determining the value of v are formidable for most planning organizations in developing countries.

In view of the severe practical problems noted above, there is little likelihood of the SCBA being used in the developing countries in the foreseeable future. The major challenge for the planning agencies is to establish a consistent framework for evaluating public investment projects and determining empirically relevant efficiency/economic prices. These prices, as shown in earlier chapters, are fairly straightforward and should not entail any major cost and staff expertise. Meanwhile, the problems of poverty and low income could be addressed through the design and implementation of projects which specifically benefit the target groups. For instance, in rural areas these would include rural development projects, rural works programs, projects for rural water supply, and health and education. However, strict economic criteria should continue to be applied to such projects to ensure that scarce public funds are not wasted in the name of poor people.

CHAPTER 13

Economic Analysis of an Irrigation Project: An Illustrative Example

Project Background

Having examined various aspects of the economic analysis of projects, we given in this and the next chapter two pracical examples of how such an analysis should be undertaken. This chapter provides a practical example of the economic analysis for an irrigation project. Let us take an imaginary country called Kohistan with a currency unit named Niglu. Kohistan is predominantly agricultural with nearly two-thirds of its gross domestic product contributed by agriculture. In the past decade, population growth has outpaced the growth in agricultural output, and the country has become a net importer of foodgrains. The government has, therefore, given very high priority to an increase in food production both to meet the growth in domestic demand and to improve the income of rural population.

At present, agricultural productivity in Kohistan is very low. The country has abundant water resources—more than adequate to irrigate the entire cultivated area. However, the water resources have, so far, not been adequately developed and only about one- fifth of the total cultivated area is irrigated. The remaining areas are entirely dependent on rain water for irrigation. Since there is virtually no scope for increasing the cultivable area, irrigation is the only way to increase Kohistan's agricultural production and productivity.

The proposed project, based on a detailed feasibility study prepared by consultants, involves the construction of a diversion weir and the development of an irrigation canal system. When fully developed, the project will help irrigate 10,000 hectares of cultivated land which at present depends entirely on rain water for irrigation. A study of the past production record made by the consultants shows that crop production in the project area has remained stagnant and no change in crop output can be expected without the project. Various alternatives for irrigating the project area were considered and the proposed project was found to be the least-cost method of achieving the objectives of the project. No new cultivated area will be opened up by the project.

The project area will draw irrigation from a large river west of the project area. The volume of water available in the river at the proposed weir site is considerably in excess of the daily needs of the proposed irrigation system throughout the year. The soil is highly suitable for the production of foodgrains like rice, wheat, and corn. The development of irrigation facilities will permit the introduction of modern technology in food production which will increase yields and cropping intensity in the project area. The boundary of the project, therefore, extends beyond the establishment and effective use of irrigation facilities. It will include an increase in food production with the adequate and timely supply of complementary inputs and the active participation of the farmers.

Financial and Economic Cost

The total financial cost of the project in local currency is estimated at Niglus (Ng) 730 million, which at the exchange rate of US$1 = Ng20 at the beginning of 1990 works out to $36.5 million. The total cost includes the base cost of Ng500 million, physical contingencies of Ng44 million, and price contingencies of Ng186 million. Details of the cost estimates are given in table 13.1. Most of the base cost is for irrigation and drainage works including the diversion weir. The balance of the cost is incurred on equipment for operation and maintenance (O&M), setting up of water user groups (WUG) and the establishment of an extension system for the entire project area, the setting up of project implementation facilities, and taxes and duties on goods imported for the project.

The financial cost includes a provision for physical contingencies equivalent to 10 per cent of the base cost for irrigation and drainage works. This provision is to allow for additional physical quantities that may be required to meet contingencies like floods, adverse soil characteristics, etc. Price contingencies,

TABLE 13.1
Financial Cost Estimates
(Ng Million)

Item	*Foreign Exchange Cost*	*Local Cost*	*Total Cost*
A. Irrigation and Drainage Works	243.0	197.0	440.0
Diversion Weir and Headworks	180.0	130.0	310.0
Irrigation and Drainage Works	60.0	55.0	115.0
On-farm Works	3.0	11.0	14.0
Land Acquisition	0	1.0	1.0
B. Equipment for O&M	5.0	1.0	6.0
C. WUG and Extension System	1.0	4.0	5.0
Civil Works	0	1.0	1.0
Vehicles and Equipment	1.0	0	1.0
Salaries and Operating Cost	0	3.0	3.0
D. Project Implementation Facilities	4.0	5.0	9.0
Civil Works	1.5	3.0	4.5
Vehicles and Equipment	2.5	0	2.5
Salaries and Operating Costs	0	1.9	1.9
Benefit Monitoring Unit	0	0.1	0.1
E. Import Taxes and Duties	0	40.0	40.0
F. Total Base Cost (1990 Prices)	253.0	247.0	500.0
G. Physical Contingencies (10% of A)	24.3	19.7	44.0
H. Price Contingencies	60.0	125.6	185.6
I. Total Financial Cost	337.3	392.3	729.6

estimated at Ng186 million, exceed one-third of the base cost of the project. This large amount is due both to the long period of construction (7 years) and provisions—based on past trends—for an annual price increase of 5 per cent for foreign exchange cost and 10 per cent for local cost.

To determine the economic cost from the financial cost given in table 13.1, two steps are required. The first step calls for the identification and exclusion of all items which do not constitute an economic cost of the project. At the same time, provisions necessary for removing any adverse impact that may be imposed on the rest of the economy should be added. In this project, no adverse impact is identified and, therefore, no additional provisions are necessary. However, three items, namely, the cost of land acquisition, import duties, and price contingencies must be excluded to arrive at the economic cost. The cost of land acquisition is excluded because it involves no real cost to the society. The economic cost in terms of lost output—which should be added to the annual flow of cost—is negligible and is, therefore, ignored. Import duties are excluded because they represent a mere transfer from the project authority to the government and, therefore, do not constitute any real cost to the economy. Price contingencies are excluded because all benefits and costs are expressed in constant 1990 (base year) prices.

The second step requires that the project cost be expressed in border prices. Since the foreign exchange cost is already in border prices (i.e., C.I.F. cost excluding import taxes), no adjustment has to be made on this cost element. To convert local costs into border prices, the following group and standard conversion factors (based on the decomposition method explained in chapter 8) have been estimated by the consultants based on information provided by the government agencies involved.

	Conversion Factors
Diversion Weir and Headworks	0.85
Irrigation and Drainage Works	0.80
On Farm and Civil Works	0.75
Equipment for O&M	0.95
Standard Conversion Factor (used for Operating Cost and Salaries)	0.90

Because of the high degree of unemployment in the project area, the real cost of hiring unskilled labor is much lower than the wage rate paid by the project authorities. Using the method described in chapter 6, the economic wage ratio in domestic prices is estimated at 0.70 and this has been used in working out the group conversion factors mentioned above. However, there is a shortage of skilled workers, for which the opportunity cost is taken to be equal to their financial wage. Using the standard conversion factor of 0.9, the economic wage ratios in border price are estimated at 0.63 for unskilled and 0.90 for skilled workers.

Using the conversion factors mentioned above, the total economic cost of the project in border prices is estimated at Ng464.8 million. Of this total, Ng277.3 million is in foreign exchange and Ng187.5 million is in local currency. Details of the economic cost under different headings are shown in table 13.2.

Based on the experience of Kohistan and in other countries at a similar stage of development, the project is estimated to be completed in seven years. The first 15 months are estimated to be spent in preparing the detailed design and engineering specifications and mobilizing the staff required to execute the project. The physical completion of irrigation and drainage works is expected to be 10 per cent in 1991, 20 per cent in 1992, 30 per cent in 1993, 20 per cent in 1994, 15 per cent in 1995, and 5 per cent in 1996. The annual phasing of the fixed economic cost, on this basis, is shown in table 13.3.

The life of the project is estimated to be 30 years after its completion in 1996. To realize the full benefits of investment, additional costs will have to be incurred over the life of the project. These include operation and maintenance (O&M) costs and expenditure on the replacement of equipment for O&M which has an economic life of only five years. The O&M cost is estimated at one per cent of the economic cost of the diversion weir and of the irrigation and drainage works (including physical contingencies), or Ng4.5 million per year beginning in 1997. The replacement cost of equipment for O&M is placed at Ng6 million every five years beginning in the year 2001. Details of these costs are shown in table 13.12.

Estimation of Project Benefits

Since the entire project area is already under cultivation, the project benefits will be measured by the increase in the value of output in 10,000 hectares brought under irrigation and net of the additional cost of inputs and labor needed for irrigated cultivation. As mentioned earlier, the consultants have already de-

TABLE 13.2
Economic Cost
(Ng Million)

Item	Foreign Exchange Cost	Local Cost	Conversion Factors	Local Cost in Border Prices	Total Cost in Border Prices
Irrigation and Drainage Works	243.0	196.0		162.8	405.8
Diversion Weir and Headworks	180.0	130.0	0.85	110.5	290.5
Canal and Drainage Works	60.0	55.0	0.80	44.0	104.0
On-farm Works	3.0	11.0	0.75	8.3	11.3
Equipment for O&M	5.0	1.0	0.95	0.9	5.9
WUG and Extension	1.0	4.0		3.5	4.5
Civil Works	0	1.0	0.75	0.8	0.8
Vehicles and Equipment	1.0	0	–	0	1.0
Salaries and Operating Costs	0	3.0	0.90	2.7	2.7
Project Implementation Facilities	4.0	5.0		4.0	8.0
Civil Works	1.5	3.0	0.75	2.2	3.7
Vehicles and Equipment	2.5	0		0	2.5
Salaries and Operating Costs	0	1.9	0.90	1.7	1.7
Benefit Monitoring Unit	0	0.1	0.90	0.1	0.1
Total Base Cost	253.0	206.0		171.2	424.2
Physical Contingencies	24.3	19.7	0.825[a]	16.3	40.6
Total Economic Cost	277.3	225.7		187.5	464.8

[a]Represents the weighted average of conversion factors under part A.

termined that crop production cannot show any change without the project. Therefore, the entire additional production is attributed to the project.

Table 13.4 gives the estimate of the area, yield, and production of various foodgrains with and without the project. As a result of the project, paddy production is estimated to go up by 51,000 tons, wheat by 16,600 tons, and maize by 1,800 tons, while a loss of 800 tons will be incurred in the production of

TABLE 13.3
Annual Phasing of Economic Cost
(Ng Million)

Item	*1990*	*1991*	*1992*	*1993*	*1994*	*1995*	*1996*	*Total*
Irrigation and Drainage	0	41.0	81.0	122.0	81.0	60.8	20.0	405.8
Equipment for O&M	0	0	0	0	0	2.0	3.9	5.9
WUG and Extension Schemes	0	0	0	0	2.0	1.5	1.0	4.5
Project Implementation Facilities	4.0	2.0	1.0	1.0	0	0	0	8.0
Base Cost	4.0	43.0	82.0	123.0	83.0	64.3	24.9	424.2
Physical Contingencies	0	4.1	8.1	12.2	8.1	6.1	2.0	40.6
Total Economic Cost	4.0	47.1	90.1	135.2	91.1	70.4	26.9	464.8

pulses which will not be grown under irrigated conditions. The increase in production is due to two interrelated factors. First, with irrigation and the application of modern inputs, yields per hectare of all crops grown will show a significant increase. The anticipated increases shown in table 13.4 are considered realistic as similar production levels have already been achieved in irrigated areas in Kohistan. Second, as a result of the availability of water throughout the year, cropping intensity is expected to go up from 140 per cent to 240 per cent.

To express the physical output mentioned in table 13.4 in monetary terms, it has to be valued in units of local currency. Since the country is a net importer of foodgrains, the economic price is equal to the cost of importing them. The import price is equal to CIF cost plus marketing and transport cost from the point of entry to major consuming centers, minus the market and transport cost involved in moving foodgrains from the project area to the urban centers.

Table 13.5 gives the estimates of economic price per ton of paddy, wheat, and maize and compares them with the financial prices prevailing in the country around mid-1990. The import

TABLE 13.4
Area, Production, and Yield With and Without the Project

Crop	With Project			Without Project			Additional
	Area (ha)	*Yield (mt/ha)*	*Output (tons)*	*Area (ha)*	*Yield (mt/ha)*	*Output (tons)*	*Output (tons)*
Cultivated Area	10,000			10,000			
Cropped Area	24,000			14,000			
Local Paddy	0	0	0	6,000	1.5	9,000	(-)9,000
Improved Paddy	16,000	4.0	64,000	2,000	2.0	4,000	60,000
Wheat	6,000	3.0	18,000	1,000	1.4	1,400	16,600
Maize	2,000	3.0	6,000	3,000	1.4	4,200	1,800
Pulses	0	0	0	2,000	0.4	800	(-)800
Cropping Intensity	(240)			(140)			

prices are those prevailing in major exporting countries in June 1990 and are considered realistic based on past trend and future projections made by international institutions. It may be noticed that except for rice, the economic price is somewhat higher than the financial price. The rice price is lower because of considerable adjustments made in import prices due to the difference in quality. In the case of pulses, which are not internationally traded, the economic price is based on its financial price, adjusted by the SCF to arrive at the border price.

Based on the increase in production estimated for different crops in table 13.4 and their economic prices derived in table 13.5, the total value of additional production is estimated to be Ng236.5 million per year (see table 13.6). This benefit will, however, be fully realized only in the year 2001, by which time the farmers are expected to reap the full advantage of the irrigation developed under the project.

Economic Cost of Inputs

Since the boundary of the project has been extended to include food production, it is essential that all costs necessary to achieve higher food output should also be included in the scope of the project. In addition to irrigation, the production of high-

TABLE 13.5
Derivation of Economic Prices of Foodgrains

Item	Wheat	Rice	Maize	Pulses
1. F.O.B. Import Price ($/mt)	140	260		100
2. F.O.B. Import Price Adjusted for Quality	140	180		100
3. Freight to Kohistan Border ($/mt)	70	60		70
4. Border Price (C.I.F.) ($/mt)[a]	210	240		170
5. Exchange Rate ($1=Ng)	20	20		20
6. Border Price (Ng/mt)	4,200	4,800		3,400
7. Freight and Marketing Cost to Major Consuming Centers (Ng/mt)	+ 400	+ 400		+ 400
8. Freight from Farmgate to Major Consuming Centers (Ng/mt)	350	380		350
9. Processing Cost of Local Grain (Ng/mt)	0	200		0
10. Farmgate Economic Price [Ng/mt (6 + 7 – 8 – 9)]	4,250	4,620 (3,000)[b]	3,350	4,500[c]
11. Farmgate Financial Price (Ng/mt)	4,000	(3,200)	2,900	5,000

[a]International freight cost to Kohistan is very high because the country is landlocked and goods have to be transshipped from the neighboring country's port.

[b]Figures in bracket represent paddy price. A recovery rate of 65 per cent from paddy to rice has been assumed.

[c]For pulses, the domestic economic price has been treated equal to the financial price. To estimate the border price, the SCF of 0.90 has been applied.

NOTE: A conversion factor of 0.75 has been used to convert domestic freight and marketing cost into border price.

yielding varieties of foodgrains require larger inputs of fertilizers, chemicals, and labor. These additional inputs must be quantified and their value deducted from the value of additional foodgrain output to arrive at the net economic benefits of the project. Estimates of the quantities of various inputs required per hectare with and without the project are shown in table 13.7.

As in the case of foodgrains, the economic price of internationally traded inputs are derived directly by estimating their CIF price and by adding to that the domestic trade and transport margins. The steps involved in estimating the economic price of fertilizers are shown in table 13.8. It will be observed that while the economic price of nitrogenous (N) fertilizer is below the financial price, the reverse is the case for phosphatic (P) and potassium (K) fertilizers.

TABLE 13.6
Valuation of Additional Output

Item	*Unit*	*Price per Unit (Ng per Ton)*	*Additional Production*	*Value of Production (Ng Million)*
Paddy	Ton	3,000	51,000	153.0
Wheat	Ton	4,250	16,600	70.5
Maize	Ton	3,350	1,800	6.0
Pulses	Ton	4,500	–800	–3.6
Crop Residue				10.6
Total				236.5

The other major inputs are seeds, chemicals, labor, and animal labor. For seeds, the economic price of each variety of grain, as calculated in table 13.5, has been applied. For the other three items, the economic prices have been derived from the financial or market prices, as shown below:

	Market Price	Conversion Factor	Economic Price
Chemicals (per ha)	500	0.9	450
Labor (per day)	20	0.9[a]	13
Animal Labor (per day)	20	0.9	18

[a]The conversion factor was applied to the shadow wage rate which was estimated at 70 per cent of the market wage rate (see page 147).

Based on the border prices of the inputs mentioned above, tables 13.9 and 13.10 provide the estimate of total input cost in the with- and without-project situations. The total cost of inputs without the project is estimated at Ng28.7 million. With the implementation of the project, the input cost will register an increase, reaching a maximum of Ng113.6 in 2001, when the full benefits of the project are expected to be realized. The net increase in the economic cost of inputs attributable to the project thus amounts to Ng84.9 million. Deducting this latter amount

TABLE 13.7
Quantities of Inputs per Hectare

Item	*Seed (kg)*	*N (kg)*	*P (kg)*	*K (kg)*	*Labor (day)*	*Animal Power (day)*
With Project						
Local Paddy	0	0	0	0	0	0
Improved Paddy	53	90	40	20	167	48
Wheat	100	100	50	20	100	21
Maize	20	60	30	30	97	25
Pulses	0	0	0	0	0	0
Without Project						
Local Paddy	55	0	0	0	120	35
Improved Paddy	52	26	8	0	150	45
Wheat	120	10	10	0	50	20
Maize	35	0	0	0	90	20
Pulses	30	0	0	0	35	5

from the increase in the value of project output, the net increase in project benefits is estimated at Ng151.6 million as shown below:

	Ng Million
1. Increase in Output	236.5
2. Increase in Inputs	84.9
3. Net Increase in Project Benefits (1–2)	151.6

Development of Irrigation Benefits

As noted earlier, the project, which started in 1990, is expected to be completed in seven years and would be fully ready by the end of 1996. However, the diversion weir and some of the irrigation canals are expected to be ready by the end of 1994 and, therefore, partial use of irrigation facilities would begin in 1995.

TABLE 13.8
Economic Price of Fertilizers

Item	*Urea*	*Triple Super Phosphate*	*Potassium Chloride*
F.O.B. Import Price ($/mt)	150	200	110
Freight to Kohistan ($/mt)	70	70	70
Border Price (C.I.F.) ($/mt)	220	270	180
Exchange Rate ($1 = Ng20)	20	20	20
Border Price (Ng/mt)	4,400	5,400	3,600
Freight to Farmgate (Ng/mt)	190	190	190
Farmgate Price (Ng/mt)	4,590	5,590	3,790
Nutrient Content	0.460	0.48	0.60
Farmgate Nutrient Price (Ng/mt)	9,980	11,645	6,317
Farmgate Financial Price (Ng/mt)	11,000	8,000	4,500

NOTE: A conversion of 0.75 has been used to convert domestic freight and marketing cost into border price.

The buildup of irrigation facilities will take four years at the rate shown below:

Year	Area Developed During		Cumulative Area	
	ha	% of total	ha	% of total
1995	2,000	20	2,000	20
1996	2,000	20	4,000	40
1997	3,000	30	7,000	70
1998	3,000	30	10,000	100

Thus, while irrigation facilities are expected to be ready by the end of 1996, their full use will be realized only in 1998.

Since the farmers in the project area are not familiar with the production technology involved in growing high-yield foodgrain varieties, a training and extension system has been built into the project. However, to be realistic, it is estimated

TABLE 13.9
Economic Valuation of Inputs With Project Implementation

Item *(1)*	*Unit* *(2)*	*Price per Unit (Ng)* *(3)*	*Input per Hectare* *(4)*	*Number of Hectares* *(5)*	*Total Input (4 x 5)* *(6)*	*Total Value of Inputs (Ng Million)* *(7)*
1. Improved Paddy						82.14
Seed	Kg	3.00	53	16,000	848,000	2.54
Nitrogen	Kg	9.98	90	16,000	1,440,000	14.37
P	Kg	11.65	40	16,000	640,000	7.45
K	Kg	6.32	20	16,000	320,000	2.02
Chemicals	Ng		450	16,000		7.20
Farm Labor	day	13	167	16,000	2,672,000	34.74
Draft Animal	day	18	48	16,000	784,000	13.82
2. Wheat						25.27
Seed	Kg	4.25	100	6,000	600,000	2.55
N	Kg	9.98	100	6,000	600,000	5.99
P	Kg	11.65	50	6,000	300,000	3.50
K	Kg	6.32	20	6,000	120,000	0.76
Chemicals	Ng		400	6,000		2.40
Farm Labor	day	13	100	6,000	600,000	7.80
Draft Animal	day	18	21	6,000	126,000	2.27
3. Maize						6.23
Seed	Kg	3.35	20	2,000	40,000	0.13
N	Kg	9.98	60	2,000	120,000	1.20
P	Kg	11.65	30	2,000	60,000	0.70
K	Kg	6.32	30	2,000	60,000	0.38
Chemicals	Ng		200	2,000		0.40
Farm Labor	day	13	97	2,000	196,000	2.52
Draft Animal	day	18	25	2,000	50,000	0.90
4. Total (1 + 2 + 3)						113.64

that it will take four years for the farmers to realize the full potentials of irrigation. The rate of growth is estimated as follows:

Year 1	30 % of full benefits
Year 2	60 % of full benefits
Year 3	80 % of full benefits
Year 4	100 % of full benefits

TABLE 13.10
Economic Valuation of Inputs Without Project Implementation

Item *(1)*	*Unit* *(2)*	*Price per Unit (Ng)* *(3)*	*Input per Hectare* *(4)*	*Number of Hectares* *(5)*	*Total Input (4 x 5)* *(6)*	*Total Value of Inputs (Ng Million)* *(7)*
Local Paddy						14.13
Seed	Kg	3.00	55	6,000	330,000	0.99
Farm Labor	day	13	120	6,000	720,000	9.36
Draft Animal	day	18	35	6,000	210,000	3.78
Improved Paddy						6.54
Seed	Kg	3.00	52	2,000	104,000	0.31
N	Kg	9.98	26	2,000	52,000	0.52
P	Kg	11.65	8	2,000	16,000	0.19
Farm Labor	day	13	150	2,000	300,000	3.90
Draft Animal	day	18	45	2,000	90,000	1.62
Wheat						1.74
Seed	Kg	4.25	120	1,000	120,000	0.51
N	Kg	9.98	10	1,000	10,000	0.10
P	Kg	11.65	10	1,000	10,000	0.12
Farm Labor	day	13	50	1,000	50,000	0.65
Draft Animal	day	18	20	1,000	20,000	0.36
Maize						4.94
Seed	Kg	3.35	35	3,000	105,000	0.35
Farm Labor	day	13	90	3,000	270,000	3.51
Draft Animal	day	18	20	3,000	60,000	1.08
Pulses						1.36
Seed	Kg	4.50	30	2,000	60,000	0.27
Farm Labor	day	13	35	2,000	70,000	0.91
Draft Animal	day	18	5	2,000	10,000	0.18
Total						28.71

Taking into account both the growth in irrigated area and the realization of irrigation potential mentioned above, the increase in net project benefits has been estimated and is shown in table 13.11. The flow of net benefits will start in 1995 and will be realized fully in 2001 when they will amount to Ng151.6 million. The net benefits are estimated to remain at that level for the entire life of the project which is expected to end in 2026.

TABLE 13.11
Growth in Net Economic Benefits from Irrigation

	1995	*1996*	*1997*	*1998*	*1999*	*2000*	*2001*
Per Cent of Total	6	18	37	63	82	94	100
Amount in Ng Million	9.1	27.3	56.1	95.5	124.3	142.5	151.6

Determination of Project Viability

Having assessed the economic cost and net economic benefits of the project, the next step is to determine its economic viability. This involves a comparison of the flow of economic cost and benefit streams. In table 13.3 the flow of fixed economic cost of the project has been estimated over the period 1990 to 1996. However, as already noted, to this has to be added the O&M cost of Ng4.5 million per year from 1997 onward, and an equipment replacement cost for O&M amounting to Ng6 million every five years beginning 2001. The annual flow of benefits beginning in 1995 to 2001 is shown in table 13.11. From that year onward the annual flow of benefits is estimated to remain unchanged at Ng151.6 million till the terminal year of the project. The residual value of the project is taken to be nil.

Since the economic life of the project is estimated at 30 years from its completion date, the flows of costs and benefits will continue till 2026. Table 13.12 gives the annual flows of costs and benefits from 1990 to the terminal year. As the flows of costs and benefits occur over different time periods, they cannot be directly compared or added up. Their present value must be determined for assessing the project's viability. The illustration uses the two main criteria used for this purpose, i.e., net present value (NPV) and economic internal rate of return (EIRR).

As already explained in chapter 9, the NPV criterion requires the determination of the opportunity cost of capital (OCC) or the minimum rate of return below which the project is not considered economically viable. Let us assume that in Kohistan the government considers that for all public sector projects the minimum rate of return should be 13 per cent. Column 5 of table 13.12 gives the present value of 1 at 13 per discount rate

TABLE 13.12
Economic Cash Flow of Irrigation Project
(Ng Million)

Year *(1)*	*Cost* *(2)*	*Net Benefits* *(3)*	*Net Flow (3–2)* *(4)*	*Discount Factor at 13 %* *(5)*	*Net Present Value (4 x 5)* *(6)*
1990	4.0	0	–4.0	1.000	–4.00
1991	47.1	0	–47.1	0.885	–41.68
1992	90.1	0	–90.1	0.783	–70.55
1993	135.2	0	–135.2	0.693	–93.69
1994	91.1	0	–91.1	0.613	–55.84
1995	70.4	9.1	–61.3	0.543	–33.28
1996	26.9	27.3	0.4	0.480	0.20
1997	4.5	56.1	51.6	0.425	21.93
1998	4.5	95.5	91.0	0.376	34.22
1999	4.5	124.3	119.8	0.333	39.89
2000	4.5	142.5	138.0	0.295	40.71
2001	10.5	151.6	141.1	0.261	36.82
2002	4.5	151.6	147.1	0.231	33.98
2003	4.5	151.6	147.1	0.204	30.01
2004	4.5	151.6	147.1	0.181	26.62
2005	4.5	151.6	147.1	0.160	23.54
2006	10.5	151.6	141.1	0.141	19.90
2007	4.5	151.6	147.1	0.125	18.39
2008	4.5	151.6	147.1	0.111	16.33
2009	4.5	151.6	147.1	0.098	14.42
2010	4.5	151.6	147.1	0.087	12.80
2011	10.5	151.6	141.1	0.077	10.86
2012	4.5	151.6	147.1	0.068	10.00
2013	4.5	151.6	147.1	0.060	8.83
2014	4.5	151.6	147.1	0.053	7.80
2015	4.5	151.6	147.1	0.047	6.91
2016	10.5	151.6	141.1	0.042	5.93
2017	4.5	151.6	147.1	0.037	5.44
2018	4.5	151.6	147.1	0.033	4.85
2019	4.5	151.6	147.1	0.029	4.27
2020	4.5	151.6	147.1	0.026	3.82
2021	10.5	151.6	141.1	0.023	3.25
2022	4.5	151.6	147.1	0.020	2.94
2023	4.5	151.6	147.1	0.018	2.65
2024	4.5	151.6	147.1	0.016	2.35
2025	4.5	151.6	147.1	0.014	2.06
2026	4.5	151.6	147.1	0.012	1.76
Total NPV 1990 to 2026					154.44

TABLE 13.13
Calculation of EIRR of Irrigation Project
(Ng Million)

Year	*Net Flow*	*NPV at 13 % Discount*	*NPV at 20 % Discount*	*NPV at 18 % Discount*
1990	–4.0	–4.00	–4.00	–4.00
1991	–47.1	–41.68	–39.25	–39.89
1992	–90.1	–70.55	–62.53	–64.69
1993	–135.2	–93.69	–78.28	–82.33
1994	–91.1	–55.84	–43.91	–47.01
1995	–61.3	–33.28	–24.64	–26.79
1996	0.4	0.20	0.13	0.15
1997	51.6	21.93	14.40	16.20
1998	91.0	34.22	21.20	24.21
1999	119.8	39.89	23.24	26.96
2000	138.0	40.71	22.36	26.45
2001	141.1	36.82	19.05	22.86
2002	147.1	33.98	16.47	20.15
2003	147.1	30.01	13.68	17.06
2004	147.1	26.62	11.46	14.56
2005	147.1	23.54	9.56	12.36
2006	141.1	19.90	7.62	10.02
2007	147.1	18.39	6.62	8.82
2008	147.1	16.33	5.59	7.50
2009	147.1	14.42	4.56	6.32
2010	147.1	12.80	3.82	5.44
2011	141.1	10.86	3.10	4.37
2012	147.1	10.00	2.65	3.82
2013	147.1	8.83	2.21	3.24
2014	147.1	7.80	1.91	2.79
2015	147.1	6.91	1.47	2.35
2016	141.1	5.93	1.27	1.98
2017	147.1	5.44	1.03	1.62
2018	147.1	4.85	0.88	1.47
2019	147.1	4.27	0.74	1.18
2020	147.1	3.82	0.59	1.03
2021	141.1	3.25	0.42	0.85
2022	147.1	2.94	0.44	0.74
2023	147.1	2.65	0.44	0.59
2024	147.1	2.35	0.30	0.59
2025	147.1	2.06	0.30	0.44
2026	147.1	1.76	0.15	0.29
Total NPV		154.44	–55.04	–18.31

for each year beginning in 1990. Applying these discount factors to the net cash flows given in column 4, the present value of the flows in each year is derived. The NPV for the entire period adds up to the positive sum of Ng154.4 million. This means that the project is highly viable because it not only satisfies the national norm of a 13 per cent rate of return but also yields a surplus of Ng154.4 million for the government.

Next, the EIRR of the project—which is the second test of project viability—is estimated. As already mentioned in chapter 9, the EIRR of a project is equal to the discount factor which makes the NPV equal to zero. Since the NPV at 13 per cent yields a large positive sum, it means that the EIRR is substantially in excess of 13 per cent. In table 13.13, the NPV is recalculated using a discount rate of 20 per cent. At this discount factor, the NPV turns out to be a negative sum of Ng55.0 million. Thus, the EIRR of the project lies between 13 and 20 per cent. To estimate the discount rate which will yield zero NPV, Ng154.4 million should be divided by the sum of the difference between the two NPVs, or Ng209.4 million. The resultant figure of 0.72 represents the distance we have to move between the two NPVs to arrive at the approximate EIRR estimate. This method yields a discount rate of about 18 per cent [13 + (0.72 x 7)]. Using this discount rate, a negative NPV of Ng18.3 million is obtained. This means that the EIRR of the project is below 18 per cent—the precise figure being 17.2 per cent.

Benefits to Farmers

The analysis in the previous section shows that the project is highly economically attractive and should be implemented. However, it is equally essential to ensure the project's financial attractiveness to the farmers who have to make use of the irrigation facilities because they will use these facilities only if they make a significant addition to their net income.

In estimating the benefits to the farmers, market prices should be used since these are the prices which the farmers pay for their inputs and receive for their outputs. The benefit is represented by the difference in the value of net output—after deducting cost of inputs—under irrigation and rainfed situation. Tables 13.14 and 13.15 give estimates of net income to the farmers under irrigation and rainfed situation, respectively, from one

TABLE 13.14
Farm Income With Project Implementation
(Ng per Hectare)

Item	*Unit*	*Improved Paddy*		*Wheat*		*Improved Maize*	
		Quantity	*Value*	*Quantity*	*Value*	*Quantity*	*Value*
1. Income							
Grain	ton	4	12,800	3	12,000	3	8,700
Crop Residue			625		–		–
Total			13,425		12,000		8,700
2. Expenditure							
Seed	kg	53	170	100	400	20	58
N	kg	90	990	100	1,100	60	660
P	kg	40	320	50	400	30	240
K	kg	20	90	20	90	20	90
Chemicals	Ng		500		445		220
Labor	day	167	3,340	100	2,000	97	1,940
Draft Animal	day	48	960	21	420	25	500
Total			6,370		4,855		3,708
3. Net Output per Hectare (1–2)			7,055		7,145		4,992
4. Cropped Area			1.6		0.6		0.2
5. Income per Crop			11,288		4,287		998
6. Total All Crops	16,573						

hectare of cultivated land. According to these estimates, the net income from one hectare of irrigated land is estimated at Ng16573, while that from rainfed land is Ng2582. Thus, the net benefit from irrigation per hectare amounts to Ng13991—nearly a fivefold increase over the without-project situation. Therefore, the project appears highly attractive to the farmers in the project area and should thus receive their full support.

As a result of the land reforms carried out by the government, most farmers in the project area own plots of land of about one hectare each. Therefore, the average financial benefit of Ng13991 per hectare fairly represents the additional income that

TABLE 13.15
Farm Income Without Project Implementation
(Ng per Hectare)

Item	Unit	Local Paddy		Improved Paddy		Wheat		Maize		Pulses	
		Qty.	*Val.*	*Qty.*	*Val.*	*Qty.*	*Val.*	*Qty.*	*Val.*	*Qty.*	*Val.*
1. Income											
Grain	ton	1.5	4,800	2.0	6,400	1.4	5,600	1.4	4,060	0.4	2,000
Crop Residue			200		300		–		–		–
Total			5,000		6,700		5,600		4,060		2,000
2. Expenditure											
Seed	kg	55	176	52	166	120	480	35	101	30	150
N	kg	–	–	26	286	10	110	–	–	–	–
P	kg	–	–	8	64	10	80	–	–	–	–
Labor	day	120	2,400	150	3,000	50	1,000	90	1,800	35	700
Draft Animal	day	35	700	45	900	20	400	20	400	5	100
Total			3,276		4,416		2,070		2,301		950
3. Net Output per hectare (1–2)			1,724		2,284		3,530		1,759		1,050
4. Cropped Area			0.6		0.20		0.1		0.3		0.2
5. Income per Crop			1,034		457		353		528		210
Total All Crops		2,582									

will accrue to the farmers in the project area. They should, therefore, be willing to pay water charges that may be levied by the government to recover at least a part of the cost of the project.

Sensitivity Analysis

The EIRR of the irrigation project has been estimated by using the most probable (but single) values of all items of costs and benefits. It is obvious that in the case of this project, many of the underlying assumptions may undergo changes which are impossible to predict at the time of the appraisal. It is, therefore, essential to measure the sensitivity of the EIRR to changes in key variables. The purpose of such an analysis is to assess the nature and extent of the risks involved, to adopt safeguards wherever possible, and more importantly, to inform policy makers on the outcome of possible adverse developments so that an informed decision could be taken about the project.

Here, the likely impact of changes in four key variables, i.e., level of output, prices of outputs, prices of inputs, and an increase in investment cost, is examined. The result of a 10 per cent change in each of these variables is shown in table 13.16. Based on the data given in this table, sensitivity indicators are estimated and shown in table 13.17. The data show that the project is quite sensitive to a reduction in the level and prices of output. For instance, a 10 per cent reduction in both variables will reduce the EIRR to less than 13 per cent and, therefore, makes it unviable. Hence, a careful look at both of them is necessary. The yield of various crops could be checked with reference to yield in other irrigated areas in the country, while prices (in constant terms) could be compared with past trends and projections made by international institutions. Our own assessment, based on past trend and future projections, suggests that the probability of these risks is quite small.

The EIRR, however, is not very sensitive to increases in input and investment costs of the project. The project, for instance, would remain viable even if input costs went up by 40 per cent or if investment costs increased by 30 per cent. Thus, the variables on the cost side are of relatively less importance than those on the benefit side of this project.

TABLE 13.16

Impact of Changes in Key Variables on EIRR of Irrigation Project

Year	Base Case	10% Lower Output	10% Lower Prices	10% Higher Input Cost	10% Increase in Investment Cost
1990	–4.0	–4.0	–4.0	–4.0	–4.4
1991	–47.1	–47.1	–47.1	–47.1	–51.8
1992	–90.1	–90.1	–90.1	- 90.1	–99.1
1993	–139.2	–135.2	–135.2	–135.2	–148.7
1994	–91.1	–91.1	–91.2	–91.2	–100.8
1995	–61.3	–63.1	–62.7	–61.8	–68.3
1996	4.0	–5.0	–3.9	–1.6	–2.3
1997	51.6	40.5	42.5	47.4	51.6
1998	91.0	72.2	76.1	83.9	91.0
1999	119.8	95.3	100.4	110.5	119.8
2000	138.0	109.9	115.7	127.4	138.0
2001	141.1	111.2	117.4	129.8	141.0
2002	147.1	117.2	123.4	135.8	147.1
2003	147.1	117.2	123.4	135.8	147.1
2004	147.1	117.2	123.4	135.8	147.1
2005	147.1	117.2	123.4	135.8	147.1
2006	141.1	111.2	117.4	129.8	141.1
2007	147.1	117.2	123.4	135.8	147.1
2008	147.1	117.2	123.4	135.8	147.1
2009	147.1	117.2	123.4	135.8	147.1
2010	147.1	117.2	123.4	135.8	147.1
2011	141.1	111.2	117.4	129.8	141.1
2012	147.1	117.2	123.4	135.8	147.1
2013	147.1	117.2	123.4	135.8	147.1
2014	147.1	117.2	123.4	135.8	147.1
2015	147.1	117.2	123.4	135.8	147.1
2016	141.1	111.2	117.4	129.8	141.1
2017	147.1	117.2	123.4	135.8	147.1
2018	147.1	117.2	123.4	135.8	147.1
2019	147.1	117.2	123.4	135.8	147.1
2020	147.1	117.2	123.4	135.8	147.1
2021	141.1	111.2	117.4	129.8	141.1
2022	147.1	117.2	123.4	135.8	147.1
2023	147.1	117.2	123.4	135.8	147.1
2024	147.1	117.2	123.4	135.8	147.1
2025	147.1	117.8	123.4	135.8	147.1
2026	147.1	117.8	123.4	135.8	147.1
EIRR	17.20	14.68	15.23	16.29	16.13

TABLE 13.17

Sensitivity Analysis of Irrigation Project

	EIRR	Sensitivity Indicator[a]
Base Case	17.2	
Output Reduced by 10%	14.7	1.5
Prices Reduced by 10%	15.2	1.2
Input Cost Reduced by 10%	16.3	0.5
Investment Cost Increased by 10%	16.1	0.6

[a]Per cent change in EIRR divided by per cent change in the variable tested.

CHAPTER 14

Economic Analysis of a Power Project: An Illustrative Example

Introduction

The previous chapter cited the example of an irrigation project in which benefits can be easily valued because most of its outputs are internationally traded. This chapter, on the other hand, will tackle a power project whose output has no market price because it is not traded in the domestic or international market. Furthermore, in most developing countries, power utilities are government-owned and tariffs levied on various consumer categories are often fixed by the government, without any direct reference to production costs or demand and supply considerations. This, therefore, raises the difficult problems of how to measure the economic benefits and assess the viability of a power project. These aspects are discussed at some length in this chapter. A brief explanation of the method of estimating the financial rate of return is also given since the financial soundness of power utilities is vital for their efficient operation and continued growth.

The hypothetical country Kohistan, which, as mentioned in chapter 13, has abundant water resources, will again be used as an example in this chapter. Kohistan's water resources have been tapped to generate power to meet the growing electricity demand. The only other domestically available resource for generating power is coal whose mining was started only a few years ago. Traditionally, wood has been the main source of noncommercial energy. However, its extensive use in the past led to large-scale deforestation and soil erosion. The high price of wood and the deliberate policy of the government to discourage its use have contributed to the rapid growth of electricity demand in Kohistan in recent years.

Because of a hilly terrain and the high cost of interconnection, three separate power grids exist in the country: one each for the eastern, western, and central region. Most of the power is consumed in the central region where the capital of Kohistan is located and where rapid population and industrial growth has been experienced. The Kohistan Electric Power Authority (KEPA) has, so far, met the entire demand for electricity in the central region through hydropower generation.

The demand for electricity in Kohistan's central region increased at an average annual rate of 12 per cent between 1980 and 1988. The GDP growth rate for that period averaged 6 per cent per year. This implies an income elasticity of demand of 2 which is much higher than in most other countries in Asia. The high growth in electricity demand is explained by the addition of many new consumers and increased demand by existing consumers. Since the process of electrification on a significant scale was started only about a decade ago, there are still many consumers—households, commercial establishments, and industrial enterprises—which are not supplied with electricity by KEPA and depend on alternative sources of energy which are both less efficient and more costly. The growth in demand for electricity is, therefore, expected to remain high in the foreseeable future.

Future Demand and Sources of Power Supply

To make a realistic estimate of the future growth in demand for electricity and to determine the least-cost option of meeting such demand, the government had, at the beginning of 1989, hired foreign consultants through technical assistance provided by an international institution. The consultants estimated that from 1989 to 1998, the demand for electricity will grow at an average annual rate of 10 per cent.

In making this estimate, four main assumptions were made: (1) GDP will continue to grow at an average annual rate of 6 per cent; (2) the government will pursue more vigorously its policy of energy conservation and industry dispersal away from the capital; (3) KEPA will install meters for all household consumers, most of whom currently pay a flat amount irrespective of the quantity of electricity consumed; and (4) the government will raise the real price of electricity by 33 per cent (from Ng1.20 to Ng1.60 per kWh), consistent with the anticipated increase in the cost of supply. These assumptions, made in close consultation with the government, are realistic and have been adopted in the analysis. The government has already raised the price of electricity from Ng1.20 per kWh in 1988 to Ng1.50 per kWh in two installments in 1989 and 1990. Another increase of Ng0.10 is proposed for 1991.

To meet the growth in power demand in the central region, the power generating capacity will have to be increased from 100 Mw at the end of 1988 to 180 Mw at the end of 1994. At the beginning of 1989, KEPA added 20 Mw of hydropower which would barely be adequate to meet the demand of the region up to the end of 1990. A balance of 60 Mw is proposed to be added during the four-year period from 1991 to 1994.

In considering the realistic options available for meeting the future demand for power from 1991 to 1994, the consultants evaluated two hydropower projects with 60 Mw capacity each, and one project involving the location of two coal-based power generation units of 30 Mw each about 20 km from the pithead of a recently opened coal mine. A comparison of these three options showed that at a discount rate of 13 per cent (representing the OCC), the setting up of two coal-based units of 30 Mw each represented the least-cost option among the three alternatives. The consultants, therefore, recommended the establishment of two coal-based units of 30 Mw each, together with associated transmission and distribution facilities. This proposal was accepted by the government and KEPA.

Expansion Program and Its Financing

In 1989 the government decided to implement the expansion program proposed by the consultants for meeting the growing demand for electricity in the central region from 1991 to 1994. This involved the setting up of two thermal generating units of 30 Mw each, the construction of a 100 km-long 220 kV transmission line, and the expansion and strengthening of distribution facilities which includes the installation of household meters. Initially, KEPA, with the government's approval, decided to implement the entire four-year investment program using the country's own resources. It placed an order with foreign suppliers for one generation unit, and for the imported materials and equipment required for the installation of a 100-km long 220 kV line. Delivery was scheduled for early 1991. The total cost of this investment program (hereafter referred to as Project I), in

TABLE 14.1
Investment in Power Project I
(Ng Million in Constant Prices of January 1990)

Item	*Foreign Exchange*	*Local Cost*	*Total*
I. Generation	240	60	300
Equipment	220	20	240
Construction	20	40	60
Materials	20	20	40
Labor	–	20	20
II. Transmission	260	80	340
Equipment	240	20	260
Construction	20	60	80
Materials	20	20	40
Labor	–	40	40
III. Tax	–	50	50
IV. Physical Contingencies	–	20	20
V. Total (I + II + III + IV)	500	210	710

constant January 1990 prices, was Ng710 million. Table 14.1 provides a breakdown of this cost under major heads.

By the end of 1990 the external resources position of Kohistan had rapidly deteriorated, and the country was obliged to seek external assistance for financing the foreign exchange cost of the second 30-Mw generating unit and for expanding and strengthening the distribution network in the central region. A project was formulated for this purpose involving a total cost of Ng720 million in constant January 1991 prices (hereafter referred to as Project II). Details of the foreign exchange and local cost of this project are shown in table 14.2. The international institution which approved the project agreed to finance the entire foreign exchange cost of Ng440 million, while the government of Kohistan and KEPA will meet the local cost of Ng280 million.

TABLE 14.2
Investment in Power Project II
(Ng Million in Constant Prices of January 1991)

Item	*Foreign Exchange*	*Local Cost*	*Total Cost*
I. Generation	260	60	320
Equipment	240	20	260
Construction	20	40	60
Materials	20	20	40
Labor	–	20	20
II. Distribution	140	140	280
Equipment	100	20	120
Construction and Installation	40	120	160
Materials	40	40	80
Labor	–	80	80
III. Tax	–	40	40
IV. Financial Contingencies	20	20	40
V. Physical Contingencies	20	20	20
VI. Total (I...V)	440	280	720

It should be noted that it is not possible to separately determine the financial or economic viability of Project I (which is funded exclusively by the government) and Project II (implemented with the help of an international institution). All components necessary to realize the full benefits of the projects, i.e., generation, transmission, and distribution, must be combined in one package to determine the overall viability of the total investment program. This means that both projects must be viewed together, and, as long as the total program is viable, the individual projects are justified because they constitute the least-cost solution in meeting projected energy demand. The financial and economic analysis of projects in the power sector should be undertaken on the basis of a time slice of the entire program rather than for individual sub-components. The analysis below, therefore, compares the aggregate costs and benefits of both projects.[1]

Investment Cost of the Power Program: 1991–1994

Financial Investment

To combine the investment cost of the two projects, they must first be expressed in the prices of a common base year, which in this case is 1991. Since the cost of Project II is already expressed in those prices, the conversion of the cost of Project I from 1990 to 1991 prices is the only other step to be undertaken. Assuming an inflation rate of 10 per cent during 1990 for both the foreign exchange and local cost components, the cost of Project I would increase to Ng781 million, as shown in table 14.3. This table also gives the cost of Project II which is the same as given in table 14.2. Financial contingencies are excluded since the financial rate of return is measured in constant 1991 prices. The total financial cost of the investment program in constant January 1991 prices is thus Ng1,461 million.

Economic Cost

To estimate the economic cost of the program, it is necessary to exclude taxes and duties and to convert the local cost to border prices. Foreign exchange costs need no adjustment as they are already expressed in border prices. The consultants had estimated the following group conversion factors for the different local cost components applied in estimating the economic cost (in border prices) of both projects:

Transport (for local cost under equipment)	0.7
Construction	0.8
Labor (EWR = MWR)	0.9

Note that since most of the labor force to be employed under the project is composed of scarce skilled workers, the economic wage rate is assumed equal to the market wage rate.

Applying the above conversion factors to the local cost components of both projects, and excluding taxes and duties which are in the nature of transfer payments, the economic cost of both projects works out to Ng1,296 million. Table 14.4 provides a summary of these costs under major heads.

TABLE 14.3
Financial Cost of Power Investment Program: 1991-1994
(Ng Million in Constant Prices of January 1991)

Item	*Project I*			*Project II*		
	FE	*Local*	*Total*	*FE*	*Local*	*Total*
Generation	264	66	330	260	60	320
Equipment	242	22	264	240	20	260
Construction	22	44	66	20	40	60
Materials	22	22	44	20	20	40
Labor	–	22	22	–	20	20
Transmission	286	88	374			
Equipment	264	22	286			
Construction	22	66	88			
Materials	22	22	44			
Labor	–	44	44			
Distribution				140	140	280
Equipment				100	20	120
Construction and Installation				40	120	160
Materials				40	40	80
Labor				–	80	80
Taxes	–	55	55	–	40	40
Physical Contingencies	–	22	22	20	20	40
Total	550	231	781	420	260	680

Phasing of Investment

The next step is to determine the combined annual phasing of the financial and economic investments of both projects. While it was assumed that Project I was initiated in 1990, the actual investment began in early 1991. Project II was approved by the international institution in 1991, but the actual procurement and installation of equipment are expected to start at the beginning of 1992. The consultants had estimated that the total time re-

TABLE 14.4
Economic Cost of Power Program
(Ng Million in Constant 1991 Prices)

	Project I			*Project II*		
Item	*FE*	*Local*	*Total*	*FE*	*Local*	*Total*
Generation	264	53	317	260	48	308
Transmission	286	72	358			
Distribution				140	118	258
Physical Contingencies	–	18	18	20	16	36
Total	550	144	694	420	182	602

quired to complete the entire investment is two years. However, it is conservatively assumed that for both projects, the time required will be two and a half years (in the ratio of 40-40-20) for completing the investment. On this basis, the combined annual phasing of the financial and economic investment cost has been estimated and is shown in table 14.5.

Benefits of Power Program

Having estimated the flow of investment cost of the power program from 1991 to 1994, the next step is to estimate the financial and economic benefits of that program. This section separately explains the estimation of both types of benefits.

Financial Benefits

Estimating the net financial benefits of the power program is relatively straightforward. It depends upon the total increase in revenues, on the one hand, and operating expenses, on the other. The revenues depend on the average tariff and sale of additional electricity. The sale of electricity, in turn, is a function of new power generating capacity, auxiliary consumption, plant load factor, and transmission and distribution losses. As mentioned earlier, the power generating capacity at the end of 1988 was 100 Mw, to which 20 Mw was added in the beginning of 1989, taking

TABLE 14.5
Annual Phasing of Financial and Economic Investment Cost of Power Program
(Ng Million In Constant 1991 Prices)

Year	*Financial*	*Economic*
1991	313	278
1992	585	521
1993	427	377
1994	136	120
Total	1,461	1,296

the total to 120 Mw. The addition of 60 Mw from 1991 to 1994 will raise the power generating capacity to 180 Mw by the end of 1994. Operating expenses have a nearly fixed relationship with the investment cost and power generation as shown in table 14.6. This table shows the estimated additional revenue from the increase in power generating capacity during 1991 to 1995.

Table 14.6 shows that revenues from additional power generating capacity depend on three main items: load factor, transmission and distribution losses, and coal consumption. The load factor, which showed a small decline from 52 per cent in 1988 to 51 per cent in 1990, is expected to register a steady increase, reaching a level of 55 per cent in 1995. The low level and the decline in the load factor is explained by two main factors. First, electricity consumption by households—the largest single consumer group—is highly uneven during the day, adversely affecting the overall load factor. Second, the flow of water in the reservoirs is low in the winter months, resulting in the operation of the power plants at substantially reduced levels.

Coal-based power plants, which can be operated at a high level of capacity utilization throughout the year, will increase the overall load factor. The increase in the relative share of industrial and commercial demand will also help improve the load factor. While the overall load factor is expected to increase from 51 per cent to 55 per cent, the load factor for the incremental supply from the two thermal plants works out to 63 per cent. The transmission and distribution losses are expected to show

TABLE 14.6
Net Revenue from Additional Power Generating Capacity from 1990 to 1995

Heading	*1988*	*1990*	*1993*	*1994*	*1995*
1. Generation Capacity (Mw)	100	120	150	180	180
2. Maximum Generating Capacity (Gwh)	876	1051	1315	1577	1577
3. Plant Load Factor (per cent)	52	51	52	53	55
4. Actual Generation (Gwh)	452	536	615	766	867
5. Auxiliary Consumption (1% for hydro and 10% for thermal)	5	5	13	28	38
6. Quantity Available for Sale (Gwh)	447	531	602	738	829
7. Transmission and Distribution Losses (per cent)	20	20	19	18	18
8. Electricity Sold (Gwh)	358	425	488	605	680
9. Increase in Sale over 1990 (Gwh)			63	180	255
10. Additional Revenue in million Ng (at Ng1.60 per kwh)			101.0	288.0	408.0
11. Operating Expenses					
i) Generation (2.5% of capital cost)			32.5	36.5	36.5
ii) Transmission (1.5% of CC)			19.5	22.0	22.0
iii) Distribution (1.5% of CC)			19.5	22.0	22.0
iv) Coal Consumption (0.65 kg per kwh of generation at Ng350 per ton)			18.0	52.0	75.0
12. Net Revenue in Ng million (10-11)			21.5	155.5	252.5

NOTE:. The estimates for actual power generation and additional power sold are made on the assumption that the first generating unit will become operational at the beginning of July 1993 while the second unit will start generating power a year later.

some reduction because of the high voltage transmission line and strengthening of the distribution network under the project. Providing all households with meters will also help reduce distribution losses and increase the revenues of KEPA.

As a result of new investments during 1991–1994, the financial revenues of KEPA are estimated to increase by Ng101 million in 1993, Ng288 million in 1994, and Ng408 million in 1995—when the full benefits from the increased electricity supply will be realized. From these revenues, operating expenses have to be deducted to arrive at the net increase in revenues realized by KEPA. The operating expenses have been estimated at 2.5 per cent of investment costs (see table 14.5) for generation and 1.5 per cent each for transmission and distribution. The largest single item of operating expenses is coal which is required to run the two thermal plants. Taking into account the relatively high ash content of coal, the consultants estimated coal consumption at 0.65 kg per kWh of power generation. This estimate is realistic and has been adopted. The cost of coal at the project site is based on the prevailing market price in Kohistan.

Deducting the operating expenses from operating revenues, the net increase in revenue is estimated at Ng21.5 million in 1993, Ng155.5 million in 1994, and Ng252.5 million in 1995. Net revenues are estimated to remain unchanged at the 1995 level throughout the remaining economic life of the program which ends in 2014. The residual value of investment estimated at Ng100 million has been treated as a benefit of the program in 2014.

Economic Benefits

In economic analysis, the point of departure is the way revenue is estimated from the increase in power sales mentioned against item 9 in table 14.6. The increase in electricity consumption is divided into two broad categories. The first category includes consumption by new consumers who will shift from alternative sources of energy to electricity generated by KEPA. In their case, resource cost saving—or saving in cost incurred on using alternative sources of energy—constitutes the benefit of the increased power supply. The second category includes the additional electricity consumed by the existing consumers serviced by KEPA. In their case, benefits are measured by what the consumers would be willing to pay for the additional electricity supplied to them.

We begin with the estimate of electricity consumption by major consumer categories in 1990 and 1995, and the estimate of the additional consumption by each category resulting from investment during the program period. Table 14.7 gives the per cent share and total consumption of each consumer group in 1990 and 1995. Notice that the share of households and the government in total consumption is expected to decline, while that of the industrial and commercial sectors will register an increase.

Based on past trends, consumers who currently depend on alternative sources of energy are expected to shift to electricity

TABLE 14.7
Share of Major Consumer Groups in Power Consumption
(Gwh)

Consumer Groups	Per cent of Total		Amount Consumed		
	1990	*1995*	*1990*	*1995*	*Increase*
Households	45	42	191	286	95
Industrial	40	43	170	292	122
Commercial	10	11	43	75	32
Government	5	4	21	27	6
Total	100	100	425	680	255

generated by KEPA. Their present alternative source of energy and consumption in kW or kWh equivalent is also indicated.

- It is estimated that an additional 8,000 households which currently depend on kerosene for lighting will start using electricity by the end of 1995. On an average, each household is assumed to consume kerosene equivalent to 120 watts of lighting, 6 hours a day throughout the year, with a fuel consumption rate of 1.25 kWh per liter. The economic price of kerosene is Ng5/liter. It is further assumed that the capital cost of lighting is roughly equal to rudimentary electrical wiring and has, therefore, been ignored.
- Some 100 industrial units will use the additional electricity generated by KEPA. They currently depend on diesel generators with an average capacity of 100 kW each and a load factor of 60 per cent. The fuel consumption rate is 3.0 kWh per liter and the economic price of diesel is Ng4.5. The capital cost of acquiring diesel sets is $600 per kW (Ng20 = $1) with the average life of the sets assumed to be 5 years.
- KEPA will provide electricity to 800 commercial establishments which currently depend on gasoline-based generating sets. The average capacity of these generating sets is 4 kW with a load factor of 50 per cent. The cost of these sets is $350 per kW and the economic price of gasoline is Ng5 per liter. The average life of these generating sets is assumed to be 4 years.

Based on the above estimates, total electricity consumption by new consumers will be 114.3 Gwh out of the total additional consumption of 255 Gwh (see table 14.8). The resource cost saving resulting from switching to electricity supplied by KEPA is estimated at Ng333.9 million. It has been assumed that the entire increase in electricity consumption by the government will come from the increased demand by existing departments and activities.

TABLE 14.8
Resource Cost Saving and Electricity Supplied to New Consumers

	Resource Cost Saving per kWh			*Electricity*	*RC*
Consumer Groups	*Fuel Cost (Ng)*	*Capital Cost (Ng)*	*Total*	*Consumed (Gwh)*	*Saving (Ng Million)*
Households	4.00	–	4.00	47.3	189.2
Industrial	1.50	0.46	1.96	52.8	103.5
Commercial	2.50	0.40	2.90	14.2	41.2
Government	–	–	–	–	–
Total				114.3	333.9

The balance of 140.7 Gwh of additional electricity supplied by KEPA from the thermal plants will be used by existing consumers. The economic price for this amount should, as mentioned earlier, be based on the price that the consumers are willing to pay for additional electricity. This price is a function of consumer demand, the alternative sources of supply, and the constraints on KEPA in meeting that demand. To simplify our analysis, it has been assumed that the existing consumers have not faced any major supply constraints and the existing tariff represents the lower bound of the willingness to pay. The tariff of Ng1.6 per kWh is, however, expressed in domestic price and has to be converted into its border price equivalent to estimate its economic benefit. Considering that the tariff rate is determined after taking into account taxes and duties and high domestic transport costs, the border cost of electricity is estimated (through the decomposition method described in chapter 8) to be 0.8 times the domestic cost, or Ng1.28 per kWh. Applying this rate to the additional amount of 140.7 Gwh of electricity consumed by existing consumers in 1995, the total benefit of additional electricity consumed by existing consumers (already given in border prices), together with the benefits of new consumers, is estimated at Ng514 million, as against the financial benefit of Ng408 million during the same year. Details of the economic benefit are given in table 14.9.

TABLE 14.9
Total Economic Benefit from Additional Sale of Electricity
(Ng Million)

	New Consumers		*Existing Consumers*		*Total Economic*
	RC Saving	*Amount*	*Tariff*	*Amount*	*Benefits*
Household Groups	*(Ng per kWh)*	*(Gwh)*	*(Ng per kWh)*	*(Gwh)*	*2 x 3 + 4 x 5*
(1)	*(2)*	*(3)*	*(4)*	*(5)*	*(6)*
Households	4.00	47.3	1.28	47.7	250.2
Industrial	1.96	52.6	1.28	69.4	191.9
Commercial	2.90	14.2	1.28	17.8	64.6
Government	–	–	1.28	6.0	7.2
Total					514.4

The above estimate is for 1995, the first year in which the full benefits of the investment program will be realized. Applying the average electricity tariff to the amount of additional electricity sold in 1993 and 1994, the revenue realized in those years is estimated at Ng127 million and Ng363 million, respectively. From these, economic benefits have to be deducted the operating costs expressed in economic prices. These operating costs are derived by multiplying the ratios of three types of operating expenses (see table 14.6) with the economic investment cost of Ng1261 million and by using the economic price of coal. The FOB export price of coal in Kohistan is $25 per ton. To estimate its economic price for domestic consumption, the transport cost from the pithead to the boarder has to be deducted, and transport cost from the pithead to the plant site has to be added. It is assumed that both costs are Ng75 per ton and, therefore, fully offset each other. The economic price of coal for use in the thermal plant is thus estimated at Ng500 per ton. With this price, the estimate of net economic revenue from 1993 to 1995 can be seen in table 14.10.

TABLE 14.10
Net Additional Revenue from the Sale of Electricity
(Ng Million)

Heading	*1993*	*1994*	*1995*
Total Revenues	127.0	363.0	514.4
Operating Expenses			
Generation	29.5	32.5	32.5
Transmission	17.5	19.5	19.5
Distribution	17.5	19.5	19.5
Coal Consumption	25.5	84.5	107.5
Net Revenue	37.0	207.0	335.4

The net economic benefit from additional power generation is thus estimated to increase from Ng37 million in 1993 to Ng335 million in 1995. The latter net benefit is expected to remain unchanged throughout the economic life of the investment program which will end in the year 2014. During that year, the residual value of the equipment measured in economic prices will have to be added to the benefit. Using the conversion factor of 0.8, the residual economic value of the equipment will be Ng80 million.

Financial and Economic Evaluation

The next step is to make a financial and economic evaluation of the investment program being undertaken by KEPA in the central region. As in chapter 13, both the NPV and EIRR criteria are used to determine the financial and economic viability of the program.

Financial Viability

In determining financial viability, it is necessary to compare the flow of financial costs and the net financial benefits of the program over its economic life (estimated at 20 years after the completion of investment). Table 14.11 gives the flow of financial

TABLE 14.11
Cash Flow of the Power Program
(Ng million)

	Financial Flow			*Economic Flow*		
Year *(1)*	*Cost* *(2)*	*Benefits* *(3)*	*Net Flow* *4 (3–2)*	*Cost* *(5)*	*Benefits* *(6)*	*Net Flow* *7 (6–5)*
1991	313	–	–313	278	–	–278
1992	585	–	–585	521	–	–521
1993	427	22	–405	377	37	–340
1994	136	165	20	120	207	87
1995	–	253	253	–	335	335
1996	–	253	253	–	335	335
1997	–	253	253	–	335	335
1998	–	253	253	–	335	335
1999	–	253	253	–	335	335
2000	–	253	253	–	335	335
2001	–	253	253	–	335	335
2002	–	253	253	–	335	335
2003	–	253	253	–	335	335
2004	–	253	253	–	335	335
2005	–	253	253	–	335	335
2006	–	253	253	–	335	335
2007	–	253	253	–	335	335
2008	–	253	253	–	335	335
2009	–	253	253	–	335	335
2010	–	253	253	–	335	335
2011	–	253	253	–	335	335
2012	–	253	253	–	335	335
2013	–	253	253	–	335	335
2014	–	353	353	–	415	415
NPV (Ng million)			79			690
FIRR/EIRR (per cent)			14.2			20.8

costs and net financial benefits which are expressed in constant 1991 prices. The flow of financial costs is taken from table 14.5, while the net financial benefits are taken from table 14.6. The net benefits are kept constant at the level achieved in 1995. In the last year of economic life of the program, Ng100 million have been added, representing the residual value of the amount invested.

It will be seen from table 14.11 that at the discount rate of 13 per cent—considered the opportunity cost of capital by the government of Kohistan—the net present value is a positive sum of Ng79 million. This means that the investment program will not only provide a 13 per cent return but will also yield an additional income of Ng79 million. Since the NPV is positive, it is obvious that the financial rate of return of the investment program (FIRR)—or the rate at which benefits (revenues) and costs are equal—will be in excess of 13 per cent. Specifically, the exact figure is 14.2 per cent. Thus, both criteria suggest that the investment program is sound and viable and should thus be implemented.

Economic Evaluation

In determining the economic viability of a project, the flows of economic costs and benefits should be compared. These data, derived from tables 14.5 and 14.10, are shown in table 14.11. The net present value at a 13 per cent discount rate translates to Ng690 million, while the EIRR is estimated at 20.8 per cent. Therefore, based on economic criteria, i.e., the contribution of the project to national output, the power investment program is highly attractive and deserves to be implemented.

The boundary of the program for both financial and economic evaluation is the same; therefore, the difference in the two returns can be explained by the fact that financial tariffs are far below those which the consumers would be willing to pay. This is so because the cost of alternative sources of energy per kWh is much higher than the tariff charged by the Kohistan Electric Power Authority. This would suggest that, if for some reason, the actual electricity sold out of the new capacity falls short of the estimated amount, there is considerable leeway for raising the financial tariff before the consumers decide to shift to alternative sources of energy.

Sensitivity Analysis and Project Risks

The method of undertaking sensitivity analysis and risk evaluation in the power-generating sector is similar to the irrigation project in chapter 13. It is, therefore, not necessary to go over the ground again. The main point to stress is that the key indicators which need to be tested for sensitivity in the power sector may be quite different from those in the irrigation sector. These need to be identified, and the impact of changes in them tested. The most important indicator to test is the supply of power from the new thermal stations. Additional supply from these plants could be lower due to several factors such as lower load factor, higher transmission and distribution losses, and poorer quality or irregular supply of coal—which may adversely affect the operation of the thermal plants. These also constitute the principal risks of the program. It is essential to ensure that the assumptions made for various indicators are realistic and safeguards are built in the investment program so that the probability of the occurrence of these risks is minimized.

Endnote

1. For the purpose of analysis, an example is taken of a small country which has only two projects in a period of four years. In larger countries, there may be numerous projects which need to be combined to determine their overall viability. In such cases, it is essential to ensure that the entire program constitutes the least-cost option of meeting the increase in power demand.

CHAPTER 15

Cost-Benefit Analysis: Restrospect and Prospect

The previous chapters attempted to explain at some length the meaning and purpose of economic analysis for public sector projects and the method for determining their economic viability. This analysis was developed primarily because (1) the public sector played a key role in investment and growth in developing countries and (2) market prices—which form the basis of private investment decisions—were highly distorted and did not serve as an appropriate basis for determining the economic viability of projects. It was felt that the use of market prices in the economic evaluation of public sector projects will not only worsen the prevailing distortions but will also lead to the misallocation of scarce resources. It was also believed that the existing distortions will continue to persist and, therefore, second-best solutions should be found through the use of "shadow" or economic prices.

In recent years serious questions have been raised about the dominant role of the state in total investment and the acceptance of the continued distortions in the market prices of various inputs and outputs. In fact, because of the persistent poor performance of public sector enterprises, many developing countries have either limited the role of the public sector or are in the process of doing so. Major adjustments in the prices of inputs, factors of production, and foreign exchange are also being implemented through reforms in macroeconomic and sectoral policies. These reforms, which have important implications for cost-benefit analysis, are discussed in this concluding chapter. The discussion will be preceded by a brief overview of the relationship between the financial and economic profitability of projects.

Financial and Economic Measure of Profitability: An Overview

It has been explained that the starting point of economic analysis is the estimate of financial costs and benefits which are based on market prices. These are the prices which both the public and private sectors face in the market place and which determine the cost of inputs purchased and the value of output sold by them. Commercial profitability based on market prices constitutes the primary basis for guiding investment decisions of individual enterprises. Profitability is generally measured by the return on total investment financed both from equity and loans. Therefore, the rate of return is calculated on the nonfinancial flows of the entity. Financial flows, however, become relevant when the return has to be calculated on owner's equity.

To estimate economic costs and benefits, several adjustments need to be made on the financial costs and benefits of a project. These adjustments essentially fall under four broad categories. First, the external impact of implementing a project must be accounted for. A project may impose additional costs or confer additional benefits which may not be reflected in financial accounts. Past experience shows that such costs are far more common than benefits. Examples of adverse external effects are pollution of a river resulting from discharges from a chemical plant, or environment deterioration caused by deforestation or the construction of a large storage reservoir. Such adverse effects have generally been downplayed in order to show the higher economic returns of projects.

Second, the financial and economic boundaries of a project may differ and this could result in a significant difference in both the costs and benefits of a project. In financial analysis, the boundary is defined by the project's output, while in economic analysis, it is defined by the project's benefits—which, in many instances, are greater than the output. All costs required to be incurred to achieve the project's benefits should be taken into account, irrespective of whether or not they are included in the financial cost. For instance, an irrigation project's objective may be to increase food production; as such, the project cost may include the full cost of building the storage reservoir and irrigation network. However, the project output may not be realized unless matching investments are made in extension, credit, storage, and marketing. The additional cost involved in these activities must be included in the project's cost to determine its economic viability. To identify economic benefits, the "with" and "without" principle is used. A project's benefits are measured by the net increase in output that will be contributed over and above the level that will prevail without the project. If a project adds to the physical supply, then the additional output is treated as the benefit of the project. If, on the other hand, a project only substitutes for an alternative source of supply, the benefit is measured by the real resources that will be released as a result of the discontinuation of an existing project. For instance, a hydroelectric project may displace an old diesel plant. In this case, diesel oil savings minus the net change in operating costs will constitute the benefit of the hydro project.

Having identified the net benefits and costs of a project, the third step is to value them in terms of economic prices. For the reasons noted earlier, market prices are not considered as an appropriate basis for economic valuation. On the cost side, economic prices are estimated by determining the impact of the resources withdrawn for use in the project on the rest of the economy. For this purpose, project inputs are divided into three groups: labor, traded goods, and nontraded goods. The economic cost of labor is based on the opportunity cost to the economy of withdrawing from other activities for use in the project. The value of traded inputs is based on their border price, i.e., export or FOB prices for exported inputs and import or CIF prices for imported inputs. The value of nontraded goods is based on market prices or on production costs, depending on whether the use of input reduces supply to the rest of the economy or if supply is met out of increased output.

Like costs, the valuation of benefits in economic analysis is quite different from that in financial analysis. Market prices are used in estimating the financial benefits in all projects, but in economic valuation varies with the nature of the project's output. If the output is a traded good, the world price serves as the basis for the valuation of a project's output. If the output is a nontraded good, the price that the consumers are willing to pay—or the marginal cost of incremental supply—should be used to estimate the benefits of the project. Economic analysis' most difficult task is perhaps to value nontraded goods and services produced and consumed by a project, a subject which has been discussed at some length in chapters 5 and 6.

The final step involves comparing all costs and benefits. In financial analysis, this is rather straightforward since market prices are used in calculating all inflows and outflows of an entity. In economic analysis, the task is complicated by the fact that some of the economic costs and benefits are expressed in domestic prices, while others are expressed in border or international prices. Therefore, a common denominator has to be used for combining them. This book uses public income (measured in foreign exchange) as the unit of account or numeraire for measuring all the benefits and costs of a project. Under this method, traded goods are valued directly at border prices while nontraded goods are valued in terms of their indirect impact on foreign exchange. The latter requires tracing down the direct or indirect foreign exchange cost of producing nontraded goods or finding traded goods for which nontraded goods are substitutes. For certain goods and services where such trading is not possible, the standard conversion factor is used. Chapter 8 explains in detail the use of conversion factors for expressing all nontraded costs and benefits in border prices.

Having determined the financial and economic value of all costs and benefits, the calculation of profitability—which forms the basis of all investment decisions—is relatively simple. Profitability is measured by the rate of return which equates the present value of the flows of costs and benefits taking place at different periods. This profitability concept is the same both in the financial and economic analysis of projects. In both cases, the higher the rate of return, the better the project. However, determining the minimum norms of profitability is different in two cases. While the financial norm is derived from the real cost of borrowed funds, economic analysis is determined by the opportunity cost of capital, or what the government can earn by using the investment funds in alternative projects.

Owing to the reasons mentioned above, the financial and economic rates of return are not strictly comparable. However, in situations where there are no significant externalities and the project boundaries are identical, the difference between the economic and financial profitability can be reasonably compared and ascribed to the distortions in the market prices of goods and services and foreign exchange. Even in cases where projects lead to external effects and the financial and economic boundaries are not identical, the price distortions are likely to be the dominant factors in explaining the differences between the economic and financial profitability of projects. Therefore, the major difference in the two measures of profitability can be reduced only through the adoption of policies aimed at removing these prevailing distortions and bringing the market prices in line with their economic values.

Implications of Public Sector Performance on the Role of State

Cost-benefit analysis was developed in the 1960s when the public sector played a dominant role in investments in most developing countries. This analysis was intended to help developing countries make economically sound investment decisions in the face of the serious price distortions that were prevailing in these countries. It was taken for granted that these distortions would continue indefinitely because developing countries would not bring about any major policy changes required to remove them. It was also assumed that the public sector enterprises would be run efficiently and that the large surplus generated would be used to hasten the process of economic development.

The actual performance record of public enterprises in the past three decades has, with a few exceptions, been highly unsatisfactory. This is due to several reasons.

First, economic analysis placed primary emphasis on the benefits of a project to the national economy and not to the project entity. Adequate attention was, until recently, not given to the recovery of cost through the proper pricing of goods and services provided by public enterprises. The issue of cost recovery is crucial because, while costs are entirely borne by the project entity in most public sector projects, the benefits accrue to the private sector. This is, for instance, true of projects in agriculture, power, transport, and the social sectors. In many countries, tariffs and fees have been so low that they do not even cover the operation and maintenance expenditure of the entity.

Second, in many cases, the prices of public sector outputs have been controlled by governments without considering the rising cost of inputs and factors of production. Governments were committed to make good the growing losses incurred by these enterprises by using budgetary resources. In actual practice, these commitments have not been met because of serious fiscal problems faced by many governments. As a result, enterprises have failed to meet growing demand for their services while the quality of their services has deteriorated in several countries.

Third, most public enterprises have experienced administrative and institutional difficulties. In many developing countries, public sector enterprises have been plagued with excessive delays in project implementation due to the slow release of funds, problems of coordination among various implementing agencies, and a shortage of qualified and experienced staff. These problems have persisted over the past three decades and have contributed to large cost overruns which have reduced the profitability of public enterprises.

Fourth, in many countries, public enterprises have been treated as a source of employment rather than profit. The excessive level of employment has resulted not only in increased operational cost but has also affected the productive efficiency of public enterprises. The absence of a system of reward for outstanding performance and penalty for inefficiency has led to a general deterioration in the overall performance of public undertakings.

The reasons mentioned above have spurred a major rethinking regarding the role of public sector in economic development and the use of cost-benefit analysis for justifying public investment. It has been increasingly realized that it is virtually impossible to effectively implement and operate projects in an unhealthy policy environment. Therefore, many developing countries have embarked upon major policy changes both at the macroeconomic and sectoral levels. The macroeconomic changes involve inter-alia realignment of exchange rates, reduced reliance on domestic and external borrowing, rationalization and reduction of indirect taxes—in particular import duties—and a reduced level of deficit financing to contain inflation. At the sectoral level, stress is placed on institutional improvements both in the implementing ministries and at the project level. At the same time, reforms are being introduced in the pricing policies of public enterprises to make them financially profitable and increasingly self reliant. In general, instead of accepting the prevailing distortions as given, deliberate efforts are being made to reduce and, wherever possible, eliminate them.

More fundamentally, serious question are being raised about the continued large-scale role of the state in total national investment. In the 1960s, the launching of large-scale public sector investment programs was justified on the grounds that the public sector would generate large-scale surpluses which would be used for speeding up the development process. These expectations have been completely belied by the actual performance of most government undertakings. Instead of yielding profits, most undertakings have incurred heavy losses, have become a drain on the exchequer, and have contributed to the large budgetary deficits experienced by many developing countries.

It is now well accepted that the public sector, because of several inherent constraints, cannot successfully compete with the private sector in terms of performance and profitability. Therefore, it should not undertake activities which can be better performed by the private sector. In particular, governments should keep out of such areas as manufacturing, tourism, banking, aviation and communications. In fact, governments should even encourage the private sector to participate in the production of electricity, coal, and other energy sources. Consequently, in the past decade, there has been a growing trend towards the privatization of the existing public sector enterprises and the reduced role of the state in new investments. This trend is expected to gain further momentum in the coming years.

Future Role of Cost-Benefit Analysis

For the reasons mentioned above, the role of the state in capital accumulation is expected to diminish greatly in the future. Henceforth, the primary responsibility for capital accumulation and growth will rest with the private sector while the state will render a supportive role by providing the necessary infrastructure facilities and social services like education and health. With a reduced, though more sharpened role of the public sector in total investment, the focus of cost-benefit analysis will also change.

Among the sectors which will remain in the domain of the state, social sector projects do not lend themselves to cost-benefit analysis. This is primarily because it is not possible to quantify their benefits and give them monetary value. Therefore, the sectors in which cost-benefit analysis will continue to be applied will be electricity and economic infrastructure including irrigation and transport. In electricity and transport, the valuation process poses some complex problems because their output is not freely traded. The benefits are measured by an estimate of the resource cost saving, or the price that the consumers are willing to pay, both of which involve judgment.

It may be noted that energy and infrastructure, which will be the principal focus of future public sector investment, are highly capital-intensive in nature. It is, therefore, essential that investment decisions and cost-benefit analysis are based on realistic sets of data. There are several areas in which improvements appear necessary to ensure that the economic benefits of public sector projects are maximized and the services provided by them are made available in the most efficient, timely, and economic manner. Some suggestions in this regard are offered below.

First, it is essential to make a realistic assessment of the future demand for government services, taking into account all key variables before determining the size of a project. Such analysis is essential to ensure that production capacities created under the project closely match the demand for them so that the project does not create either excess capacities—which will lead to waste of scarce public funds—or shortages that will adversely affect the productive activities of the private sector.

Second, a careful examination should be made of the various options available for delivering a given service; the option which involves the least cost should be selected. For instance, a project may involve the movement of a given quantity of goods which could be done by railways, road transport, or a combination of the two. All the three options should be carefully examined and it should not be too readily assumed that such options do not exist.

Third, tariffs for government services like electricity and transport should be at levels which not only cover operation and maintenance costs but also meet a good part of the cost of future expansion required in those sectors. In the past, tariffs had been kept low in the mistaken belief that these would promote growth in productive sectors. In actual practice, governments have not been able to properly maintain existing services and provide resources for further development because of low tariffs. As a result, scarcities have emerged and these have become a major impediment to growth.

Fourth, since electricity and transport are natural monopolies, the tariffs charged for such services do not serve as an adequate basis for judging the operational efficiency of public enterprises. Some objective norms must be prescribed to ensure that these undertakings are run efficiently and at least cost to consumers.

Fifth, many public sector projects have failed to yield the expected returns because of institutional deficiencies in project implementation and operation. In the past, it was too readily assumed that required institutional support will be forthcoming. Since institutional aspects have a major bearing on the economic viability of projects, they should be made an integral part of the cost-benefit analysis.

Sixth, although cost-benefit analysis considers the environmental impact in project appraisal, its actual use has remained limited on the plea that it is difficult to accurately measure it in monetary terms. With the growing environmental concern worldwide, a more thorough assessment will have to be made of the environmental impact of projects and, as much as possible, the cost of environmental protection should be treated as a part of total project cost.

In conclusion, the focus of cost-benefit analysis will have to change with the reduced role of the state in capital accumulation. The principal areas in which this analysis will continue to be applied are irrigation, power, and transport, in which the role of governments will remain predominant. Most of these areas are highly capital-intensive and, therefore, efficiency and economy in the use of public funds will have to be assured. At the same time, since all these services are vital for private sector growth, care will have to be taken to ensure that these services grow in line with the needs of the private sector. The project analyst will also have to work more closely with other experts involved in project appraisal to ensure that those aspects which have a crucial bearing on the determination of project profitability are estimated on a realistic basis.

In the electricity and transport sectors, the absence of market prices has led tariffs to be based on the financial cost of providing service to the consumers. In general, economic benefits measured by resource cost savings and the consumers' willingness to pay are greater than the revenue from tariffs. This represents the upper limit to which the tariffs could be raised. With the narrowing of the gap between market and economic prices, the difference between financial and economic profitability may disappear over time. This will even be more true if the project boundaries are the same and the external impact of projects is internalized in estimating their financial profitability. It is, however, unlikely that all market distortions will entirely disappear in the years to come. Therefore, cost-benefit analysis will continue to remain as an indispensable tool in the selection and implementation of public sector projects.

The economic efficiency criteria will continue to be the principal basis for determining the economic profitability of projects. With the reduced importance of public sector investments, social cost-benefit analysis will become even less relevant because the public sector cannot be expected to carry the burden of bringing about distributional changes in the economy when its role is confined to a few sectors. The primary role of the project analyst will not be to attempt social cost-benefit analysis but to conduct a more refined economic analysis in which existing tools are used more efficiently and effectively in improving the economic and financial performance of public sector projects.

Appendix

Discount Rate Table
(Present value of 1 at rate *i* payable in *t* years)

n/i	*2%*	*3%*	*4%*	*5%*	*6%*	*7%*
1	0.980	0.971	0.962	0.952	0.943	0.935
2	0.961	0.943	0.925	0.907	0.890	0.873
3	0.942	0.915	0.889	0.864	0.840	0.816
4	0.924	0.888	0.855	0.823	0.792	0.763
5	0.906	0.863	0.822	0.784	0.747	0.713
6	0.888	0.837	0.790	0.746	0.705	0.666
7	0.871	0.813	0.760	0.711	0.665	0.623
8	0.853	0.789	0.731	0.677	0.627	0.582
9	0.837	0.766	0.703	0.645	0.592	0.544
10	0.820	0.744	0.676	0.614	0.558	0.508
11	0.804	0.722	0.650	0.585	0.527	0.475
12	0.788	0.701	0.625	0.557	0.497	0.444
13	0.773	0.681	0.601	0.530	0.469	0.415
14	0.758	0.661	0.577	0.505	0.442	0.388
15	0.743	0.642	0.555	0.481	0.417	0.362
16	0.728	0.623	0.534	0.458	0.394	0.339
17	0.714	0.605	0.513	0.436	0.371	0.317
18	0.700	0.587	0.494	0.416	0.350	0.296
19	0.686	0.570	0.475	0.396	0.331	0.277
20	0.673	0.554	0.456	0.377	0.312	0.258
21	0.660	0.538	0.439	0.359	0.294	0.242
22	0.647	0.522	0.422	0.342	0.278	0.226
23	0.634	0.507	0.406	0.326	0.262	0.211
24	0.622	0.492	0.390	0.310	0.247	0.197
25	0.610	0.478	0.375	0.295	0.233	0.184
26	0.598	0.464	0.361	0.281	0.220	0.172
27	0.586	0.450	0.347	0.268	0.207	0.161
28	0.574	0.437	0.333	0.255	0.196	0.150
29	0.563	0.424	0.321	0.243	0.185	0.141
30	0.552	0.412	0.308	0.231	0.174	0.131
40	0.453	0.307	0.208	0.142	0.097	0.067
50	0.372	0.228	0.141	0.087	0.054	0.034

Continued on next page

Discount Rate Table (*Continued*)
(Present value of 1 at rate *i* payable in *t* years)

n/i	*8%*	*9%*	*10%*	*11%*	*12%*	*13%*	*14%*
1	0.926	0.917	0.909	0.901	0.893	0.885	0.877
2	0.857	0.842	8.826	0.812	0.797	0.783	0.769
3	0.794	0.772	0.751	0.731	0.712	0.693	0.675
4	0.735	0.708	0.683	0.659	0.636	0.613	0.592
5	0.681	0.650	0.621	0.593	0.567	0.543	0.519
6	0.630	0.596	0.564	0.535	0.507	0.480	0.456
7	0.583	0.547	0.513	0.482	0.452	0.425	0.400
8	0.540	0.502	0.467	0.434	0.404	0.376	0.351
9	0.500	0.460	0.424	0.391	0.361	0.333	0.308
10	0.463	0.422	0.386	0.352	0.322	0.295	0.270
11	0.429	0.388	0.350	0.317	0.287	0.261	0.237
12	0.397	0.356	0.319	0.286	0.257	0.231	0.208
13	0.368	0.326	0.290	0.258	0.229	0.204	0.182
14	0.340	0.299	0.263	0.232	0.205	0.181	0.160
15	0.315	0.275	0.239	0.209	0.183	0.160	0.140
16	0.292	0.252	0.218	0.188	0.163	0.141	0.123
17	0.270	0.231	0.198	0.170	0.146	0.125	0.108
18	0.250	0.212	0.180	0.153	0.130	0.111	0.095
19	0.232	0.194	0.164	0.138	0.116	0.098	0.083
20	0.215	0.178	0.149	0.124	0.104	0.087	0.073
21	0.199	0.164	0.135	0.112	0.093	0.077	0.064
22	0.184	0.150	0.123	0.101	0.083	0.068	0.056
23	0.170	0.138	0.112	0.091	0.074	0.060	0.049
24	0.158	0.126	0.102	0.082	0.066	0.053	0.043
25	0.146	0.116	0.092	0.074	0.059	0.047	0.038
26	0.135	0.106	0.084	0.066	0.053	0.042	0.033
27	0.125	0.098	0.076	0.060	0.047	0.037	0.029
28	0.116	0.090	0.069	0.054	0.042	0.033	0.026
29	0.107	0.082	0.063	0.048	0.037	0.029	0.022
30	0.099	0.075	0.057	0.044	0.033	0.026	0.020
40	0.046	0.032	0.022	0.015	0.011	0.008	0.005
50	0.021	0.013	0.009	0.005	0.003	0.002	0.001

Continued on next page

Discount Rate Table (*Continued*)
(Present value of 1 at rate *i* Payable in *t* years)

n/i	*15%*	*16%*	*18%*	*20%*	*25%*	*30%*	*40%*
1	0.870	0.862	0.847	0.833	0.800	0.769	0.714
2	0.756	0.743	0.718	0.694	0.640	0.592	0.510
3	0.658	0.641	0.609	0.579	0.512	0.455	0.364
4	0.572	0.552	0.516	0.482	0.410	0.350	0.260
5	0.497	0.476	0.437	0.402	0.328	0.269	0.186
6	0.432	0.410	0.370	0.335	0.262	0.207	0.133
7	0.376	0.354	0.314	0.279	0.210	0.159	0.095
8	0.327	0.305	0.266	0.233	0.168	0.123	0.068
9	0.284	0.263	0.225	0.194	0.134	0.094	0.048
10	0.247	0.227	0.191	0.162	0.107	0.073	0.036
11	0.215	0.195	0.162	0.135	0.086	0.056	0.025
12	0.187	0.168	0.137	0.112	0.069	0.043	0.018
13	0.163	0.145	0.116	0.093	0.055	0.033	0.013
14	0.141	0.125	0.099	0.078	0.044	0.025	0.009
15	0.123	0.108	0.084	0.065	0.035	0.020	0.006
16	0.107	0.093	0.071	0.054	0.028	0.015	0.005
17	0.093	0.080	0.060	0.045	0.023	0.012	0.003
18	0.081	0.069	0.051	0.038	0.018	0.009	0.002
19	0.070	0.060	0.043	0.031	0.014	0.007	0.002
20	0.061	0.051	0.037	0.026	0.012	0.005	0.001
21	0.053	0.044	0.031	0.022	0.009	0.004	0.001
22	0.046	0.038	0.026	0.018	0.007	0.003	0.001
23	0.040	0.033	0.022	0.015	0.006	0.002	
24	0.035	0.028	0.019	0.013	0.005	0.002	
25	0.030	0.024	0.016	0.010	0.004	0.001	
26	0.026	0.021	0.014	0.009	0.003	0.001	
27	0.023	0.018	0.011	0.007	0.002	0.001	
28	0.020	0.016	0.010	0.006	0.002	0.001	
29	0.017	0.014	0.008	0.005	0.002		
30	0.015	0.012	0.007	0.004	0.001		
40	0.004	0.003	0.001	0.001			
50	0.001	0.001					

References

Ali, Ifzal. 1986. *Public Investment Criteria: Economic Internal Rate of Return and Equalizing Discount Rate.* Report Series, no. 37. Manila: Asian Development Bank.

_______. 1989. *A Framework for Evaluating Economic Benefits of Power Projects.* Staff Paper, no. 43. Manila: Asian Development Bank.

_______. 1990. *Public Investment Criteria: Financial and Economic Internal Rates of Return.* Report Series, no. 50. Manila: Asian Development Bank.

Anderson, Lee G., and Russel F. Settle. 1977. *Benefit-Cost Analysis: A Practical Guide.* Lexington, Massachusetts: Lexington Books.

Asian Development Bank. 1987. *Guidelines for Economic Analysis of Projects.* Manila.

Baum, Warren C., and Stokes M. Tolbert. 1985. "Investing in Development: Lessons of World Bank Experience". *Finance and Development*, vol. 22, no. 4 (December).

Barish, N., and S. Kalpan. 1978. *Economic Analysis for Engineering and Managerial Decision Making.* 2nd ed. New York: McGraw-Hill.

Gutierrez-Santos, L.E., and G. Westley. 1979. *Economic Analysis of Electricity Supply.* Washington, D.C.: Inter-American Development Bank.

Harberger, Arnold C. 1976. *Project Evaluation: Collected Papers.* Chicago: The University of Chicago Press.

Irwin, George. 1978. *Modern Cost-Benefits Methods: An Introduction to Financial, Economic and Social Appraisal of Development Projects.* London: The MacMillan Press Ltd.

Levin, Herman M. 1983. *Cost Effectiveness: A Primer.* Beverly Hills, California: Sage Publications.

Little, I.M.D., and J.A. Mirrlees. 1974. *Project Appraisal and Planning in Developing Countries.* New York: Basic Books.

________. 1990. "Project Appraisal and Planning Twenty Years On". Paper Presented at the World Bank Annual Conference on Developing Economies, World Bank, Washington, D.C.

Marglin, Stephen A. 1967. *Public Investment Criteria: Benefit-Cost Analysis in Planned Economic Growth.* Cambridge, Mass.: The M.I.T. Press.

Pearce, W.D., and C.A. Nash. 1981. *The Social Appraisal of Projects: A Textbook in Cost-Benefit Analysis.* New York: John Wiley and Company.

Pouliquen, Louis Y. 1970. *Risk Analysis in Project Appraisal.* World Bank Staff Occasional Paper, no. 11. Baltimore: The Johns Hopkins University Press.

Ray, Anandrup. 1984. *Cost-Benefit Analysis: Issues and Methodologies.* Baltimore: The Johns Hopkins University Press.

Roemer, Michael, and Joseph J. Stern. 1974. *The Appraisal of Development Projects: A Practical Guide to Project Analysis with Case Studies and Solutions.* New York: Praeger Publications.

Schiender, Hartmut. 1978. *National Objectives and Project Appraisal in Developing Countries.* Paris: Organisation for Economic Cooperation and Development Centre.

Schwartz, Hugh, and Richard Berney, eds. 1977. *Social and Economic Dimensions of Project Evaluation.* Washington, D.C.: Inter-American Development Bank.

Squire, Lyn, and Herman G. van der Tak. 1975. *Economic Analysis of Projects.* Baltimore: The Johns Hopkins University Press.

United Nations Industrial Development Organization. 1974. *Guidelines for Project Evaluation.* New York.

Ward, W.A., and B.J. Deven. 1988. "A Practical Guide to Economic Analysis of Agricultural Projects." Washington, D.C.: World Bank, Economic Development Institute. Mimeographed.

Webb, M., and D. Pearce. 1984. "The Economic Value of Power Supply." Washington, D.C.: World Bank. Mimeographed.